NOTHING WITHOUT TOURISM

ANTHROPOLOGY OF EUROPE

General Editors:
Monica Heintz, University of Paris Nanterre
Patrick Heady, Max Planck Institute for Social Anthropology

Europe, a region characterized by its diversity and speed of change, is the latest area to attract current anthropological research and scholarship that challenges the prevailing views of classical anthropology. Situated at the frontier of the social sciences and humanities, the anthropology of Europe is born out of traditional ethnology, anthropology, folklore and cultural studies, but engages in innovative interdisciplinary approaches.

Anthropology of Europe publishes fieldwork monographs by young and established scholars, as well as edited volumes on particular regions or aspects of European society. The series pays special attention to studies with a strong comparative component, addressing theoretical questions of interest to both anthropologists and other scholars working in related fields.

Volume 10
Nothing Without Tourism
Local Dependency, Time and Nativism in the Swiss Alps
Danaé Leitenberg

Volume 9
The Everyday Politics of Food Co-ops
Care, Aid and Community in Austerity Britain
Celia Plender

Volume 8
Translocal Care across Kosovo's Borders
Reconfiguring Kinship along Gender and Generational Lines
Carolin Leutloff-Grandits

Volume 7
Gentrifications
Views from Europe
Marie Chabrol, Anaïs Collet, Matthieu Giroud, Lydie Launay, Max Rousseau and Hovig Ter Minassian

Volume 6
A Taste for Oppression
A Political Ethnography of Everyday Life in Belarus
Ronan Hervouet

Volume 5
Punks and Skins United
Identity, Class and the Economics of an Eastern German Subculture
Aimar Ventsel

Volume 4
In Pursuit of Belonging
Forging an Ethical Life in European-Turkish Spaces
Susan Beth Rottmann

Volume 3
All or None
Cooperation and Sustainability in Italy's Red Belt
Alison Sánchez Hall

Volume 2
European Anthropologies
Edited by Andrés Barrera-González, Monica Heintz and Anna Horolets

Volume 1
The France of the Little-Middles
A Suburban Housing Development in Greater Paris
Marie Cartier, Isabelle Coutant, Olivier Masclet, and Yasmine Siblot

NOTHING WITHOUT TOURISM

Local Dependency, Time and Nativism in the Swiss Alps

Danaé Leitenberg

berghahn
NEW YORK • OXFORD
www.berghahnbooks.com

First published in 2026 by
Berghahn Books
www.berghahnbooks.com

Library of Congress Cataloging-in-Publication Data

Names: Leitenberg, Danaé author
Title: Nothing without tourism : local dependency, time and nativism in the Swiss Alps / Danaé Leitenberg.
Description: First edition. | New York : Berghahn Books, 2026. | Series: Anthropology of Europe ; volume 10 | Includes bibliographical references and index.
Identifiers: LCCN 2025026572 (print) | LCCN 2025026573 (ebook) | ISBN 9781836951728 hardback | ISBN 9781836951711 epub | ISBN 9781836951704 adobe pdf
Subjects: LCSH: Community-based tourism--Switzerland--Bernese Alps | Nativism--Switzerland--Bernese Alps | Ethnology--Switzerland | Globalization and tourism
Classification: LCC G156.5.C65 L45 2026 (print) | LCC G156.5.C65 (ebook) | DDC 910.9494/7--dc23/eng/20250821
LC record available at https://lccn.loc.gov/2025026572
LC ebook record available at https://lccn.loc.gov/2025026573

British Library Cataloguing in Publication Data

A catalogue record for this book is available from the British Library

EU GPSR Authorized Representative

LOGOS EUROPE, 9 rue Nicolas Poussin, 17000, LA ROCHELLE, France
Email: Contact@logoseurope.eu

ISBN 978-1-83695-172-8 hardback
ISBN 978-1-83695-171-1 epub
ISBN 978-1-83695-170-4 web pdf

https://doi.org/10.3167/ 9781836951728

CONTENTS

ACKNOWLEDGEMENTS

First and foremost, I extend my deepest gratitude to all those I met during my field research. Their trust, openness and willingness to share their life stories, their fears, hopes, joys and struggles, made this work possible. I can only hope that I have honoured their voices as faithfully as possible in these pages.

I am profoundly grateful to my supervisor and mentor, Sabine Strasser, at the Institute of Social Anthropology at the University of Berne, and to my colleague and friend Paul Reade. Their guidance and encouragement within the project *Trapped in Paradise: Entangled Mobilities and Imaginaries of Freedom*, funded by the Swiss National Science Foundation (SNSF), provided the foundation upon which the core ideas of this book took shape. Sabine, your trust and tireless intellectual, administrative and emotional support – from fieldwork to writing up – have been invaluable. Your dedication to making Swiss academia a better place is deeply inspiring. Paul, thank you for the laughter, the thought-provoking conversations, and the countless ways in which you have enriched this journey. I am also immensely grateful to Valerio Simoni, my second supervisor, for his thoughtful feedback and invaluable insights during my PhD.

A heartfelt thank you to my colleague and friend Annika Lems, whose guidance and intellectual companionship have accompanied me throughout my career – from my early research days in Berne to our time in Germany, within the research group *Alpine Histories of Global Change* that she led at the Max Planck Institute for Social Anthropology in Halle (Saale). Our ongoing exchanges, and those with my project colleagues Christine Moderbacher and Markus Wurzer, have profoundly shaped my understanding of the Alps.

I am especially grateful to my dear friend Kiri Santer for her unwavering intellectual and personal support, through desperate phone calls, motivational speeches, critical feedback, and shared laughter over the years.

Thank you to Andrea Boscoboinik, Andrea Friedli, Angela Lindt, Andrew Haxby, Gerhild Perl, Ghassan Hage, Herta Nöbauer, Ikuno Naka, Jacqueline Kalbermatter, Janina Kehr, Jelena Tošić, Jeremy Rayner, Johannes Truffer, Julia Rehsmann, Julia Vorhölter, Manuel Insberg, Mario Schmidt, Mirela Ivanova, Oliver Nachtwey, Rahel Jud, Sabrina Stallone, Siqi Tu, Sonja Moghaddari, Veronika Siegl, Viviane Cretton and Zeynep Sarıaslan for their invaluable insights and encouragement. I also extend my sincere appreciation to the Swiss National Science Foundation, the Dr. Joséphine de Kármán Foundation, and the Max Planck Society for their financial support.

My heartfelt thanks to Tom Bonnington at Berghahn Books for his editorial guidance, and to the two anonymous reviewers for their invaluable feedback on this manuscript.

To my dear friends Anthony, Camille, Céline D. and G., Fabia, Lea, Marsali – thank you for your emotional and intellectual support throughout the years of writing this book. A special thank you to my family – my wonderful parents Beatrix and Jacques, my amazing sister Naomi, and my inspiring grandmother Ruth – for your unconditional love and care throughout my whole life, especially during those times when I felt like giving up. And finally, to Artan, who has walked this journey with me like no one else has, thank you for your patience, love and care.

PRELIMINARY NOTE

Given the contentious processes, events and ideologies that shape the small Alpine village I study, I have chosen to anonymize the name of my field site in this book. To protect the community and its activities, I refer to the location as 'the village', 'the valley', 'the municipality', 'the resort' or the 'field site', using these terms interchangeably or to reflect different contexts (geographical, historical, social or legal). This approach to location anonymization is more superficial, and thus imperfect, than the deep anonymization strategy I have used for individuals: it has evolved with time, alongside my research findings and publications, and occurs within a highly publicized and promoted field of tourism in the Jungfrau Region and the Bernese Highlands. It is further constrained by certain obligations, particularly the need to acknowledge the names of entities whose work I reference, in order to comply with copyright laws governing the use of protected material under 'fair use' (notably in chapters 2 and 3). This anonymization strategy is the result of my efforts to balance my duty to protect informants, my obligation to appropriately cite material, and my task as a researcher to 'provide a meaningful analysis of "real" issues, people and places' (Vorhölter 2021: 16).

All names of individuals involved in this book have been anonymized in order to protect my informants, as is common practice in social anthropology (Pels et al. 2018; de Koning et al. 2019). I have adopted a conventional anonymization strategy in qualitative research by making people unrecognizable, to both community members and the reader (Meyermann and Porzelt 2014). To do so, I have changed their names, genders, attributes, ages and nationalities, and have merged claims and divided them. In some cases, this anonymity is fragile and relative, as people have taken part in extensive public and media coverage. In such sensitive instances, I have weighed in the actors' wishes to tell 'their side of the story' and my duty as a researcher not to contribute to further harm (Saunders, Kitzinger and Kitzinger 2015).

References

de Koning, Martijn, et al. 2019. 'Guidelines for Anthropological Research: Data Management, Ethics and Integrity', *Ethnography* 20(2): 170–74.

Meyermann, Alexia, and Maike Porzelt. 2014. 'Hinweise zur Anonymisierung von qualitativen Daten', *Forschungsdaten Bildung Informiert* 1.

Pels, Peter, et al. 2018. 'Data Management in Anthropology: The Next Phase in Ethics Governance?' *Social Anthropology* 26(3): 391–413.

Saunders, Benjamin, Jenny Kitzinger and Celia Kitzinger. 2015. 'Anonymising Interview Data: Challenges and Compromise in Practice', *Qualitative Research* 15(5): 616–32.

Vorhölter, Julia. 2021. 'Anthropology Anonymous? Pseudonyms and Confidentiality as Challenges for Ethnography in the Twenty-First Century', *Ethnoscripts* 23(1): 15–33.

INTRODUCTION

After the train meanders through the forest, crossing bridges and busy roads, all passengers turn to the right with an ecstatic 'Wow' at the dramatic sight of the *Dreigestirn* mountains. Hands grab at phones and cameras to capture the view that so many have painted, written and sung about since the Romantic origins of tourism in the late eighteenth century. The giant wall of ice and rock seems to have erupted overnight from the green pastures, as if the mighty mountains had awoken and suddenly sat up, taking the whole valley by surprise. Twice an hour, crowds comprising mostly tourists reach the train station of the Alpine village, approximately 900 metres above sea level, located in the Jungfrau Region in the Bernese Swiss Alps.

From summer 2017 until spring 2023, I often belonged to that crowd too, taking the train from my familiar and urban Swiss lowlands (*Unterland*) where I had lived and studied for most of my life to reach the not-so-distant but oh-so-different highlands (*Oberland*), where my field site, a village of four thousand permanent residents, was located.[1] Together with hundreds of people from hundreds of places, speaking hundreds of languages, I would undertake a journey that crossed what often felt like an invisible border between the ordinary and the extraordinary, between the insignificant Swiss flatlands and the exceptional mountains. State-of-the-art transport infrastructure supported our smooth arrival into, as well as our movement within, the valley. Multilingual advertisements scattered through the village centre promised us unforgettable mountain experiences and assured that we had chosen to come to the right place. Signs in German, English, French, Arabic or Korean hanging on the façades of restaurants and hotels, in front of sports and souvenir shops, and of boutiques selling luxury clothing and furniture, all competed for our attention as guests, inviting us to come in and spend as much time and money as possible.

And it seemed to work: each day of my stay, thousands of people moved up and down the main village street, the *Hauptstrasse*, primarily along the one-kilometre stretch between the train station and the church – a small yet bustling corridor within the nearly 200-square-kilometre expanse of the valley. Pedestrians along the pavements paused to capture photos of the wilder surroundings, beyond the urban centre, where a panorama, made up of mountain peaks, pastures, huts and grazing cows, promised an Alpine experience of escape. Beside them, more people navigated this elongated centre using cars, buses, coaches, vans, bicycles and motorcycles, often with the intent to go higher and further to enjoy the views and outdoor activities like hiking, skydiving, sledging and skiing, depending on the weather and the season. Many visited the village's shops, hotels and restaurants, browsed tourist information desks, and boarded gondolas in facilities conveniently lined up along both sides of the street. The multilingual murmur of the crowd, the hum of tyres, the screech of brakes, the honking of horns, and the announcements of train arrivals and departures from the country's lowlands – all seemed to account for the valley's success as an international tourist destination, a place where the so-called 'business of foreigners' or *Fremdenverkehr*, a German synonym for tourism, dominated everything.

* * *

After days spent around the busy *Hauptstrasse*, I became increasingly interested in one of the many hotels bordering it. I had noticed how the paint peeled off its Victorian façade, and traces of damp mapped out strange motifs on its walls. The light garland framing the main entrance to the lobby flickered eerily and the red flowers decorating its windows hung down their pots. What could have been negligence was accompanied by a disturbing quietness in the middle of one of Switzerland's most touristic villages. After quick online research, I read that the hotel had been empty for years, waiting for an investor. My father, who visited me, insisted that I should 'find out' what had happened to it, as it was 'terrible for the image' of the resort.

As I questioned locals on the state of the hotel, they highlighted its 'dramatic' fate – 'a jewel', 'a five star' that had become a 'ruin' and was awaiting an investor that many doubted would ever come. But they also worryingly told me about other 'ruins', all bordering the *Hauptstrasse*. Farther up, another empty hotel caused concerns as it had been left half-demolished. What should have been fully renovated was still unfinished. With only one façade demolished and the majority of the interior destroyed, the money was already gone. The village's tourism association had paid to cover the ruins and debris with a mural reproducing the original appearance of the building – to make it look 'less frightening', I was told. Locals also mentioned a giant orange crane that stood in a square, concrete hole in the centre of the village. What should have been a luxurious hotel had turned into a never-ending construction site. The crane moved from time to time, but many suspected that this was just to give passers-by the feeling that things were going according to plan. Locals remembered that there was a time in the recent past when seven hotel projects were awaiting the helping hand of investors.

These abandoned or unfinished projects particularly stood out because they did not match with my expectations of what I would find in this area of the Alps. Before coming to the village to conduct fieldwork in September 2017, I had read the overwhelmingly enthusiastic statements of tourism experts from the regional, national and international media on how the valley was blessed with rugged views that boosted an exceptionally successful tourism economy. Thanks to the ongoing attraction of the valley's mountains and the relentless work of tourism lobbyists (*Touristikers*) in attracting new markets and extending the tourist high season, the local tourism economy was praised for its robustness and expansion. Such narratives surely matched with what Salazar and Graburn have termed 'tourism imaginaries': an assemblage of preconceived notions made of 'seductive images and discourses about peoples and places' (Salazar and Graburn 2014: 1) – in this case, of a peaceful Alpine village. But they were also based on impressive figures. In the year I arrived, the number of visitors had boomed, with approximately twenty thousand day trippers and 1.2 million overnight stays annually, breaking all records in the lengthy local history of tourism. And in terms of visitor numbers, the next two years, 2018 and 2019, were expected to surpass the previous one. The Covid-19 pandemic hit the village in 2020, but only two years later, in 2022, the resort had once again grown and done better than ever.

Trying to 'find out' what had happened to these hotels, I found myself trying to understand how to reconcile the obvious successes of a tourism economy that had contributed to the expansion and development of the valley over centuries, despite its blatant failures and uncertainties. Talking to tourists proved rather inconclusive in this respect: many of those I talked to had not noticed the 'ruins', or simply did not care. The strategies employed by local tourism actors, such as using a mural or light garlands to deflect attention, appeared to have been effective. But evoking the ruins with those who lived off tourism in the valley, all four thousand inhabitants – more or less directly, as I was told – opened up larger, often deeply existential, conversations on tourism's ambiguity and their position as dependents of this industry.

My encounter with Adrian, a local in his fifties who worked in one of the 'ruins', was particularly memorable in this respect. In charge of keeping one of the empty hotels in a condition that would be likely to attract new investment, he was, to the best of his knowledge, the only villager of local 'native' origin or *Einheimische* who had spent most of his life in the hotel business. His former colleagues were nearly all migrants from Southern or Eastern Europe who carried out outsourced activities typically avoided by local natives and Swiss people. Adrian, however, explained that he used to feel at home in the hotel, among guests and workers from all over the world, and his openness and skills as an electrician made him a particularly valuable worker. Now, he was the only one left in what many called the 'ghost hotel'.

On the day I met him, Adrian was dismantling the heating system and interrupted his work to give me a tour of the ruined hotel. Pointing out the old-fashioned décor in some of the remaining rooms, he spoke of the damage inflicted by the passage of time, along with the need for maintenance as well as the necessity to reinvent oneself in the business of tourism. The rubble before our eyes was the result of opportunities that his former boss had missed, failing to adapt to a fast-changing market. Tourism was based on people coming from the other side of the world. Nowadays, tourists, especially the 'mass' or 'group tourists' from Asia, had less time and money to spend in the village, Adrian said as we walked around the half-destroyed rooms of the so-called returning guests (*Stammgäste*), who used to stay for weeks, sometimes months, and for whose needs and tastes the rooms had been tailored. The competition between various destinations and hotels had also intensified with the advent of online reviews and numerous booking options.

Adrian regretted that his former boss did not have it in him to foresee trends or to inject hundreds of thousands of Swiss francs into securing the hotel's future. This lack of innovation, this lack of focus in an accelerating economy, had led the place to bankruptcy, and left employees in shock. Everyone had to leave, except for Adrian. His colleagues, nearly all migrants, had left the village or Switzerland altogether after the bankruptcy, sometimes returning to where they had come from. Some found a job in one of the fifty hotels spread throughout the valley. Adrian had heard from one or two but the rest had vanished into thin air, it seemed, joining the hundreds, probably thousands, of migrant employees who had contributed to the hotel's history since its construction at the end of the nineteenth century and then disappeared.

Influenced by his proximity to ruins that he understood to be in direct correlation with his former boss's failure to innovate, Adrian positioned himself in favour of innovation as the only possible way forward in a touristic Alpine valley. For instance, like many locals, he enthusiastically supported a new cable-car project that aimed to secure the valley's future, even if this meant accepting that the more stable and 'good times' of tourism with the presence of the *Stammgäste* were over. But Adrian also knew that many of his acquaintances, friends and family members openly struggled with the consequences of the endless need for growth imposed on them by a global tourism economy.

In my discussions with various villagers, they indeed referred to the empty hotels – places some of them had never set foot in – but were not calling for actions or opportunities to invest further. Instead, they evoked them as symbols of the valley's misguided tourism trajectory, lamenting what had been lost in the pursuit of touristic expansion: Alpine traditions, landscapes, a sense of community and stability, but also some of the local touristic history, with the disappearance of once-prestigious hotels like the one Adrian worked in. They complained about the ever-extending tourist season, as the previously quieter months of April and May, and September to November, were now busy too. Relying on day trippers from far away was deeply bothersome, tiring and upsetting, they said. The newest forms of tourism – shorter, from farther away, and dominated by big tourism players – took over all aspects of local life and exposed the region to more global instability. At the same time, many felt that fewer and fewer 'native locals' (*Einheimische*) 'profited from tourism' as rents had increased, local shops had disappeared in favour of sports shops, and holiday rentals and big malls had been built in an increasingly urbanized Alpine valley.

If the resort's inhabitants had disagreements on the direction of tourism, they nonetheless unanimously agreed that the valley relied entirely, more or less directly, on the 'business of foreigners' to exist. For the past two centuries, modern tourism had put the valley, the Jungfrau Region, the Bernese Highlands and the Alps on the global map. Farmers, lift operators, construction workers and service providers all now counted on the presence of guests from all over the world to make a living. The 'business of foreigners' had made history, enabling locals to work and to stay in the village when other Alpine dwellers had had to emigrate to the urban lowlands to make a living. It had also brought workers from abroad to the valley, who were able to escape poverty and instability in their own countries to work in the Alps, building tourist infrastructure and serving guests.

The more I learnt about ruined, abandoned, or 'ghost' hotels, the less they seemed at odds with the resort's thriving economy. They instead illuminated what became this book's topic: the ambiguities of tourism for those who rely on it to make a living. The history of these buildings reflected the broader history of the valley, one that tourism dependents often described as a testament to tourism's remarkable ability to create and transform durably. This was especially evident in an Alpine valley that had once grappled with poverty and outmigration before the advent of tourism two centuries ago. At the same time, the hotels' more recent fates highlighted the general pressure exerted by the tourism industry on locals, dramatically underscoring their need to keep pace with the international market. If tourism shaped all aspects of life in the village, present and past, it also threw it and its inhabitants into a problematic, uncertain future that could undo everything. What did it mean, then, to live off something that seemed as essential as uncertain, as promising as threatening?

Based on eighteen months of fieldwork conducted in the valley between 2017 and 2023 for my doctoral and postdoctoral research, this book addresses this question by examining the paradoxes of tourism – considering it not only as an industry and infrastructure but also as a social and existential phenomenon. It explores how tourism provides direction, meaning and dignity to human life while simultaneously fostering confusion, disorientation and anxiety for those who depend on it for their livelihood. Told from a region that appears to exemplify tourism's robustness, the narrative in the following chapters uncovers the profound instability of tourism as a capitalist endeavour in a globalized world. It seeks to illuminate the complex experiences of tourism dependency for the inhabitants of a valley with no viable alternative to tourism, as well as the strategies they employ to navigate the promises and threats that tourism entails. To do so, I examine the perspectives of various actors within the valley's tourism industry, focusing on the stratified and hierarchized categories of belonging they occupy: those like Adrian, who identify as 'natives' or 'original locals' (*Einheimische* or *Ureinheimische*, literally meaning those 'at home' in the village), alongside other Swiss locals, migrants – primarily from Southern and Eastern Europe – working in the hospitality sector, and tourism 'makers' or suppliers (e.g. lobbyists, company executives or hoteliers) who actively shape the industry. Together, these positions reveal how being dependent on tourism involves unevenly navigating between a long history of touristification and an uncertain future, oscillating between success and ruin.

'NOTHING WITHOUT TOURISM'

Living off tourism in the Bernese Alps means being deeply connected to a long-standing and changing history of tourism development, with this industry playing a central role in shaping the region's environmental, economic and social landscape. Locals repeatedly told me that they would not exist had the tourism industry not conquered their valley, and that they were 'nothing without tourism'.[2] What was once a valley shaped by scattered settlements (*Streusiedlung*), composed of isolated farmhouses with little interaction between inhabitants from different sides and flanks, had been transformed into a galaxy revolving around the *Hauptstrasse* (main road or road), pulsating with life. The gaps between the centre and the rest of the *Streusiedlung* had gradually, yet swiftly, been filled with new constructions, particularly holiday homes, each seeking a piece of the tranquillity and pure Alpine experience that the village offered. The face of the valley had undoubtedly changed.

At the same time, in a region where the history of tourism could be traced back to the origins of modern tourism itself, with the Romantics, or, for instance, the first Thomas Cook tours, visiting the valley (Barton 2008), tourism was also a story of economic transformation. Books, exhibitions and theatre plays described how the village's present was the undeniable result of a development that shifted from mountain farming to tourism (see Chapter 1). Tourism had progressively replaced mountain farming, the latter being seen as a precarious activity undertaken in a harsh climate like in many other Alpine valleys (Stotten, Schermer and Wilson 2021), so harsh that many locals left the valley for other regions, countries and continents (Lerch 1909).[3] While 60–70 per cent of local inhabitants engaged in farming activities in the village in 1900, this figure dropped to fewer than 5 per cent five years before the beginning of fieldwork, a reality for various valleys of the Bernese Highlands (Berwert and de Bary 2005; Haisch, Jakob and Mayer 2014; Tiefenbach 2013). Many locals thus gave up on farming to become more or less directly involved in the tourism business. They found employment with transport companies that emerged in the valley from the turn of the twentieth century – most notably the private regional operator, Jungfrau Railway (JR). Others also built houses designed to welcome tourists in rental flats. Those who remained farmers in the valley managed to do so because of the cheese and meat they sold to hotels, their employment as mountain guides or on ski lifts during the quieter winter months and their reliance on state subsidies.[4]

The 'business of foreigners' grew to dictate the valley's time, its space and the relationships between inhabitants and the world as the Alpine village transformed into a 'resort' – a destination offering a curated blend of material, cultural and economic resources, thoughtfully designed to meet the needs and fulfil the desires of tourists (Clivaz et al. 2014). This was the result of a long process, made from decades of investments by private and public actors and institutions, often from the urbanized Swiss lowlands, to boost the creation of the Alpine village as a place dedicated to leisure (Tissot 2004).[5] These pushed all-inclusive group tourism from Europe at the end of the nineteenth century, and then ski tourism from the 1930s (Denning 2014). In the 1960s, Swiss ski tourists from urban areas such as Basel, Zurich and Berne often bought holiday homes in the village. From the end of the 1980s, the village increasingly welcomed guests from the United States, then Japan, and more recently China, Korea and India (Gyimóthy 2018), often travelling in organized groups on tours throughout Europe and Switzerland. Highly popular on social media, the valley had also become the centre of multiple tourism buzzes, attracting users and influencers from all over the world.

Perhaps counterintuitively, tourism's significance also manifested in its lack of clarity. Locals and even tourism officials I met during fieldwork explained that figures and facts, while providing important indications, could not truly capture the all-pervasive importance of tourism.[6] The resort had to be able to withstand four to five times the size of the municipality's population during the ever-extending high season. Yet, the exact numbers of tourists coming into the valley and what they were doing and seeking exactly remained unknown because of the complexity of this crowd, made up of day trippers, holiday-home owners and returning guests, coming in by train, bus, car, bike or even helicopter. Official statistics dividing the occupation percentages into sectors (75 per cent in the third sector, 15 per cent in the second, and 9 per cent in the first, as of 2020)[7] did little to paint a more precise picture. Farming and construction, as providers of infrastructure and products, were indivisible parts of the tourism industry.

An acknowledgement of tourism's influence in both the past and the present, the sentence 'We are nothing without tourism' also highlighted this industry's power to make people and places into *someone* or *something* in an Alpine region that would otherwise be nothing. During my field research, locals explained that their village was one of the few municipalities in the Alps not to draw on a negative demographic growth – a reality that my research interlocutors believed to be owed to tourism's successes.[8] Locals expressed that they lived in a touristic enclave located in a mountain region marked by a remoteness and a lack of opportunities and political power, as opposed to the urban centres in the country's lowlands. Tourism provided them with possibilities to make a living in the valley, when many of their neighbours struggled. They highlighted the unevenness of the Alpine development, with some places literally bursting with tourists while others were bypassed and disregarded, left out of important economic and sociocultural networks (Stacul 2003a; Bätzing 2015). As such, tourism enabled continuity in maintaining otherwise disappearing rural structures and populations – a well-known fact in the Alps and beyond (Wanner, Pröbstl-Haider and Feilhammer 2021).

Yet, with tourism as their sole viable option in an otherwise marginalized mountainous region, many villagers expressed a sense of indebtedness to tourism that carried a pessimistic undertone. Already precarious in the 1980s, most of my local interlocutors explained that the situation had worsened since the turn of the twenty-first century as they felt that the tourism industry had become increasingly alienated from the needs of the village community.[9] In a world of increasing interconnection and uncertainty, the village had entered a new reality, competing with rival destinations and trying to attract unpredictable guests from further away, notably Asia and the Middle East, instead of the European *Stammgäste*. Processes of globalization and digitalization had amplified local concerns, as they were accompanied by the growing and visible dominance of transnational corporations, such as tour operators, hotel chains and cruise companies. Being of considerable size and influence, villagers complained that these companies were putting the valley under strong pressure as they pursued their own profit – for instance, by offering group discounts in order to attract more guests.

These developments were not entirely dictated by outsiders. The private railway company JR owned trains but also restaurants, ski lifts, gondolas, sightseeing spots, shops, car parks and so on.[10] Variously, JR was described as the architect of the region's success, a 'lighthouse' (*Leuchtturm*), a 'milk cow' (*Milchkuh*) for the whole region and the country, and as one of Switzerland's oldest and most thriving railway companies from its foundation at the end of the nineteenth century. This transportation company figured as *the* most important agent of tourism development because it had pulled the village out of Alpine farming and poverty, and brought it into modernity. Yet, over the last forty years, JR had bought up multiple, publicly owned touristic installations and become a holding company, the main employer of the local workforce in the region, as well as an important player in the national economy of tourism. Amid feelings of loyalty and scepticism, locals felt more or less compelled to speak either against or for the monopoly of this historical company over the region and its recent transformations, especially regarding the future of growth and of the more intense and rapid global connections that the company promoted.

The notion of being 'nothing without tourism' encapsulates the historical, social and economic significance of tourism while also highlighting its threats as a dominating industry in the Alps. However, as I will demonstrate in this book, tourism is far from a predetermined or static phenomenon, even in a valley with such a long history of touristification. Its dominance must be continuously remade, reaffirmed, and actualized – performed and shaped by a diverse array of actors, each driven by unique goals and interests.

THE TOURED ONES

Each year, more travellers engage in tourism, be it in search of new emotional experiences (Picard and Robinson 2012), to find love (Frohlick 2012) or to connect to their roots (de Santana Pinho 2008), while an increasing number of people rely on tourism to make a living.[11] Together, they participate in the creation of what anthropologist Edward Bruner (Bruner 1996) has described as 'borderzone': 'a creative space' (ibid.: 159) and a 'zone of interaction between natives, tourists, and ethnographers' (ibid.: 177) in which people create meaning across cultural difference (Bruner 2005; Leite, Castañeda and Adams 2019; Simoni 2019).

It could be argued that tourists and myself as an ethnographer depended on tourism in the village's borderzone, either to take a break away from the heat, to rest or to work. Yet, it is a more straightforward and acute form of tourism dependency that interests me here – one informed by individuals' particular awareness of their heavy reliance on the industry for survival, growth, development, and meaningful lives.[12] This focus is informed by what the great majority of my interlocutors in the village described of the clear role they occupied in the tourism industry: theirs as the 'travelled' or 'toured ones' (*die Bereisten*), in contrast to those of the 'travelling ones' (*die Reisenden*) who visited the valley. The latter temporarily left their everyday lives to discover the valley, paying for services, food and accommodation while away from home, while the former relied on the latter's presence in their village to make a living. Some – usually the most precarious, migrant workers – even expressed that they lived not just *off* tourism but also *for* it, stressing their complete dedication to responding to the tourists' needs in order to make ends meet.

As an ethnographer of tourism, who stayed in the village for a whole year, I was identified by many as (and felt like) belonging somewhere in between the toured and the travelling ones. I was no regular tourist who would leave after staying in a hotel and enjoying a break from work because I was here *for* work, but my capacity to make a living did not directly depend on the tourism industry. From this intermediary position – shaped by practical constraints of access I will develop below, and guided by my interests in tourism's paradoxes – I chose to tell the stories of those who rely on tourism as a means of *re-creating* (rather than merely *recreating*) their everyday lives. Doing so, I focus on a particular role and research object within the heterogeneous tourism economy: the so-called 'hosts', as opposed to the 'guests'.

Anthropologist of tourism Valene Smith famously defined the tourist as 'a temporarily leisured person who voluntarily visits a place away from home for the purpose of experiencing a change' (Smith 1989: 1). Smith, like other analysts of tourism, reasserted the boundary between the visiting tourist and the local or 'native', as a permanently rooted and working inhabitant of a place that they did not necessarily choose to live in (Bruner 2005: 15; 1996). In the last decades, the host and guest dichotomy, as well as the fixed typologies of tourism's participants,[13] have been critiqued at great length in academia for their inadequacy in the face of a fluctuating and more complex reality, where one might for instance be tourist *and* worker (e.g. Sherlock 2001). This also rings true for my field site, where tourists sometimes settled permanently or where locals enjoyed their leisure time in the valley in the same way as tourists would do. Yet, I here take the separation between hosts and guests as a point of departure that is anchored in the people's perception of their dependency on tourism as an industry commodifying the extraordinary to make the everyday and a living for those inhabiting the resort. I am in this sense committed to addressing what Jenny Chio has termed the 'labour' put behind someone's 'pleasure' in tourism (Chio 2014: vii). The labour opportunities provided by tourism enable the inhabitants of resorts to make a living, and tourism can only be maintained by their continuous involvement in it. In the Alpine village, as in many touristic resorts, the commodification of extraordinary time for guests represents the dominant vehicle of social reproduction.

The way I conceive it, the 'borderzone' is no exceptional space of 'staged relations', nor is it removed from the everyday, as was originally implied in Bruner's definition. For him, after the encounter in the borderzone, 'tourists return to their hotels and the locals to their homes and families, each to their own spaces' (Bruner 2005: 251). Instead, and inspired by Valerio Simoni's rereading of the concept (Simoni 2019), I see a borderzone not solely as a 'site of work and cash income' (Bruner 2005: 267) but as an integral part of the everyday from the perspective of locals. The idea of being 'nothing without tourism' precisely reveals how even attempting to delimit the contours and limits of tourism and of that borderzone was difficult, if not impossible, because of its all-pervasiveness. For those whose stories appear here, tourism creates and is created by the everyday: the borderzone extends into their homes, their past, present and future, and even their bodies and minds.

Considering the hosts' experiences also does not equate with a focus on the 'natives', as Bruner's characterization of the touristic borderzone may imply (Bruner 1996, 2005). Like many global resorts, the valley was characterized by the great variety of its approximately four thousand inhabitants. Some were indeed 'natives' or *Einheimische* like Adrian. They carried in their names their belonging to their homeland (*Einheimische*, literally translating to 'in the homeland' or *Heimat* in German), often for multiple generations. This native status was marked by family history, a local family name, speaking a local dialect, having the village municipality inscribed as one's 'place of origin' (*Heimatort*). The *Einheimische* were seen and often felt like 'authentic' village dwellers, who originally came from the valley, making them both hosts and natives. Others were 'outsiders' (*Auswärtige*), coming from other Swiss valleys or regions, or even other countries for those who worked in various sectors of the local economy, especially in the hospitality industry. In the local everyday talk, these international migrants were addressed as 'foreigners' (*Ausländer*), and their contribution to the realm of tourism labour was seen as very situated and, to a certain extent, limited to the hospitality sector as waiters, kitchen staff or general hotel workers in the many hotels, restaurants and rentals. As of 2017, the booklet I was given as a new inhabitant in the municipality indicated that the locals who had lived in it since birth made up 31 per cent of the total village population and the proportion of foreigners was 24 per cent.[14]

Such heterogeneity of population should not be seen as exceptional. Various scholars of tourism have discussed at great length how tourism borderzones present a wide array of actors and positions that make the tourism industry (Chambers 1997) and the 'host community' (Stronza 2005), particularly as tourism resorts often rely on a migrant workforce and other 'brokers' or intermediaries (Bianchi 2000). However, we know surprisingly little about this heterogeneity; and the question of how inclusion and exclusion dynamics may unfold between different groups remains rather unexplored in full-length ethnographic studies (see Adams 1992; Tucker 2003). My work sheds light on the importance of acknowledging how such uneven categories of population are remade in the everyday to understand tourism dependency, its benefits and constraints, and its uneven manifestations.

TOURISM DEPENDENCY

Over the last decade, social scientists have increasingly described tourism as an extractive industry because of its reliance on the exploitation of environments, landscapes and people (Büscher and Davidov 2013; Azcárate 2020). And where it contributes to the creation of monocultural environments, tourism is particularly prone to shocks and crises that disrupt not only economic branches but also entire communities (Milano, Novelli and Russo 2024). In spite of this acute awareness of tourism's negative consequences, many of its dependents continue to seek opportunities to be or remain involved in tourism, as a sector that is undoubtedly tied to promises of development, stability and reparation (Mostafanezhad, Azcárate and Norum 2021).

Accounting for the paradoxes of tourism as a simultaneously creative and destructive industry (Enzensberger 1996: 121–26; Stasch 2017), many analysts have carefully distanced themselves from earlier positions describing tourism either as an agent of chaos or as a promise of betterment (Krippendorf 1975). Described by various authors as not intellectually productive and too narrow (van den Berghe 1992: 235; Loloum 2015: 37), these analyses overlooked tourism's inherent contradictions. They did not acknowledge the fact that tourism creates exotic, pure, wild and paradisiac destinations while domesticating them and subjecting them to capitalist processes of value extraction. The current momentum nonetheless underscores the inevitability of debates over tourism's impacts. In a world of social inequality, of economic and ecological insecurity or sanitary uncertainty, the nefarious and salutary effects of tourism for the environment, for local communities and for the economy have increasingly shaped academic discussion (Duffy 2006; Mostafanezhad and Hannam 2016; Moore 2019), as well as various actors from the broader society and private and public institutions.[15]

If tourism presents a case of 'moral reasoning' for those who are observing it (Caton 2012), for those who depend on it to make a living, it embodies both an existential promise and a threat. The tourism dependents I met during my fieldwork saw themselves tied to tourism as an unmanageable but vital business, relying on the unpredictable wanderlust of others. Travel for leisure was a luxury that many people would give up on first in times of crisis or inflation – especially for a trip to expensive Switzerland, as I was told multiple times. Yet, the growing crowds of tourists also weighed heavily on locals, who regularly expressed that tourism had become 'too much' as it was destructive of their natural and social environment. Both the disappearance and the presence of guests, and the past and the future of tourism development, were reasoned with intensively, revealing an acute problematization of locals' tourism dependency that echoed with other destinations globally, as we shall see.

Tourism dependency is not a new topic in tourism studies. In the 1970s and 1980s, the followers of the so-called dependency theory[16] saw the growing tourism industry in the Global South as paradigmatic of a problematic state of dependency between the powerful West and the powerless Rest (Turner 1976; Erisman 1983; Morgan 1987). Tourism, they argued, was one of the forces sustaining the underdevelopment of the former colonies or the peripheries while benefiting the industrialized core or centre (Britton 1982; for a review, see Chaperon and Bramwell 2013; Bianchi 2018). These patterns were considered even more problematic on islands or in isolated enclaves, which were often made dependent on the Western world and exhibited continuities with the plantation economy model shaped by colonization (Bastin 1984), newly driven by transnational corporations (Britton 1980). Anthropologists influenced by this scholarship, for instance, denounced the increasing reliance of indigenous populations on Western tourists visiting their home villages (Nash 1989), and feared for the disappearing native traditions, which should be salvaged from commodification (Greenwood 1989). If criticized for their essentializing portrayal of dependent local communities and determinism, such grand narratives – often formulated in terms of 'impacts' – effectively countered those of liberal optimists, who presented tourism as a positive developmental strategy enabling poorer countries to quickly and smoothly enter globalization (Simoni 2018: 508–9; Leite, Castañeda and Adams 2019).[17] As a consequence of its simplifying narratives, the dependency theory in tourism and beyond reached an 'impasse' in the late 1980s (Schuurman 1993).[18] So much so that speaking of tourism dependency became extremely marginal, especially within anthropological scholarship on tourism.

Informed by the limitations of the 'classical' tourism dependency framework, my work revisits tourism dependency debates as essential issues shaping my field site – much like other destinations – but approaches them through a different lens. It engages with recent anthropological scholarship on dependency, which has contributed to important discussions on personhood, the state or distributive politics by challenging the often simplistic and deterministic views of dependency as the negative, pathologized counterpart to freedom, individualism or choice (Martin and Yanagisako 2020; Martin, Wig, and Yanagisako 2021; Martin 2021). The sole position of my Alpine field site, as a village relying on the 'business of foreigners', points to the 'normality' that relations of dependency constitute: village inhabitants have depended on the wanderlust of others to make a living and tourists have depended on the services offered by the inhabitants to enjoy their visits. Native inhabitants of the valley have also relied on the presence of other (outsider or foreign) inhabitants migrating into the valley to sustain and develop this industry, while these 'newcomers' have depended on the valley's job opportunities to secure livelihoods.

Acknowledging this situation, I embrace an understanding of tourism dependency as a relational 'mode of action', capable of accounting for the complex history, activities and relations that mark states of dependence (Bayart 2000: 218; Ferguson 2013, 2015). In this understanding, being dependent does not indicate passivity, but instead highlights people's capacity to access resources and support and to make claims on what or whom they depend.[19] Being dependent on tourism means to possess acute awareness of tourism as a source of income and exchange, to actively reflect on it and on one's position in relation to it. Looking at the relations that shape tourism subsistence reveals how those who live off tourism formulate demands on the tourism industry, its participants and representatives, and the affective states that can accompany these demands, such as hope, fear and confusion. Sometimes, however, an empirically observable reality of tourism dependency may not lead to a socially recognized status of tourism dependent, as this book shows in the case of exploited migrant hospitality workers (see Chapter 5).

Although not framed within the context of tourism dependency, several anthropologists have continued to explore tourism's paradoxical power to open up opportunities for dependents to connect, learn and move, while also constraining them (Tucker 2005; Kahn 2011; Chio 2014; Leite 2017). Matilde Córdoba Azcárate's recent book *Stuck with Tourism* (2020) most poignantly reveals the necessity to reopen the question of tourism's ambiguity in its ability to better *and* worsen (non-)human life. Based in some of Mexico's famous touristic resorts, her research looks at how tourism 'sticks' and 'entraps' those who depend on it, precisely because it enables the 'short-term' betterment of 'precarious conditions', while foreclosing alternatives (Azcárate 2020: 16). The precarious context that Azcárate witnesses in the Yucatán Peninsula is the result of colonial processes, the decline of the henequen monocrop industry, and ruthless capitalist investment driven by the Mexican state from the mid-1970s. Her work contrasts with that of the simplistic analyses of the dependency theorists of tourism but marks a regained interest in anthropology, in fact even a necessity, in examining the states of dependence created by tourism in marginalized, postcolonial regions.

The context that informs Azcárate's insightful analysis is arguably quite different from the one that interests me here. The Swiss Alps are located in the centre of Europe, in a country that is known for its wealth and stability, and has largely profited from its neighbours' colonial enterprise (Purtschert, Lüthi and Falk 2012). Contrary to the Yucatán, my field site, like various resorts in the Swiss Alps, draws on a very long history of tourism – two hundred years at least. Yet, here too, tourism enabled locals to be 'something' in a region that would otherwise be 'nothing' – meaning struggling with poverty and the outmigration of its inhabitants to wealthier, urban and more diversified regions. Beyond their obvious differences, tourism developments and the dependency this creates for those who live off it in the Yucatán Peninsula and the Swiss Alps are marked by their greater or lesser proximity to becoming 'nothing' *again*, because of the context of marginalization and/or the precarity that characterizes their pre-tourism history.[20]

Although Azcárate reveals the strategies put in place by locals who had no other choice but to sacrifice for tourism, her book reveals that most of her interlocutors had little power to change their conditions. Local populations in the Yucatán Peninsula, she notes, resorted to various narratives to 'make sense' by seeing tourism effects as 'necessary evil' (Azcárate 2020: 8). The sacrificial logics of tourism were also present for locals who, like Adrian, argued that innovations, as painful they were, needed to be undertaken for the sake of the community. They however coexisted with more upfront problematizations of tourism dependency. For many of the people I follow in this book, tourism no longer 'made sense' because it no longer seemed to be benefiting for those it ought to – local tourism dependents. Putting forward their status as tourism dependents, as locals or citizens of Switzerland, they wished to address this situation. Such problematization of tourism by its dependents, by virtue of their dependent status as 'affected' population (see Chapter 5), reveals precisely what anthropologists of dependency recently argued: looking behind the imaginaries of subjugation associated with it, one might find that being dependent and being socially recognized as such may be marked by a greater form of agency (Ferguson 2015).

The critiques uttered by locals about tourism reflect growing, wider social movements targeted against tourism's excesses in various (especially Western) destinations. In cities like Paris, Amsterdam and Barcelona, locals have expressed discontent with tourism's excesses; they claim their right to define its development according to their needs, and call for the implementation of various regulations to safeguard their quality of life (Milano 2018; Peeters et al. 2018). In the sociological and anthropological tourism literature, these have been described as social mobilizations or activism against the phenomenon of 'overtourism' or 'mass tourism'. This strand of scholarship has discussed how locals on overtouristified destinations more or less openly confront, attack, resist or 'cope' with those associated with tourism growth, be they promoters, planners, policymakers or tourists (Boissevain 1996; Milano, Novelli and Cheer 2019; Milano, Novelli and Russo 2024). My work draws inspiration from these studies, recognizing the greater agency of Swiss and native locals in shaping, navigating and responding to their tourism dependency, in contrast to cases like those described by Azcárate. While not as influential as the 'suppliers' of tourism – such as lobbyists, experts, hoteliers and business owners who control the means of production and shape tourism to suit their interests – many locals mobilized a critique that contrasted 'quantity' tourism with 'quality' tourism. Based on this, they argued for business models that would be closer to those of the more sustainable and ethical 'slow tourism' or 'eco-tourism' (Ars and Bohanec 2010; Sidali and de Obeso 2017) instead of the alienating and accelerating 'mass tourism' imposed on them.

However, the sole mobilization of a progressive 'resistance' framework to analyse these local critiques and movements can be misleading when it fails to recognize their role in reproducing nativist, exclusionary attitudes. Contrary to urban contexts like Barcelona, where movements against overtourism stemmed from activist circles already involved in providing affordable housing or fair work (Milano, Novelli and Russo 2024: 18, Santos, Malet Calvo and Nofre 2025), the reactions against overtourism in the Swiss Alps did not draw on the same political genealogy. In a political landscape dominated by economic liberalism and social conservatism, many of the locals I met attributed tourism's nefarious consequences specifically to non-European 'mass' tourists, whom they framed as the 'bad' tourists because they deteriorated the local landscape – contrary to historical, Western tourists (see also Breiby, Øian and Aas 2021). At the same time, the migrant workers, although dependent on tourism, were largely excluded from local efforts to counter overtourism and shape its development in a 'profitable' and meaningful way for the dependent local community.

Based on these developments in the field of tourism studies and the anthropology of dependency, this book will address tourism dependency as an experience, as a problematic state of those who stand between tourism's promises and threats in a context shaped by the increasing pressure of touristic capitalism. The Alpine valley, as a place that seems to exemplify the robustness of this industry, is not exempt from this global phenomenon. At the same time, I develop an understanding of the actions, relationships and politics that shape the formation of the 'tourism dependent' category, particularly when it is mobilized to critique excessive developments.

CHRONOPOLITICS OF TOURISM, SLIPPERINESS AND ENTRAPMENT

While many scholars of tourism have emphasized the spatial dimensions of tourism development and its impacts on the livelihoods of local communities and environments (Williams and Lew 2014; Chen and Chen 2017; Azcárate 2020), the problematization of tourism dependency I encountered during fieldwork mainly manifested as a temporal concern. Individually and collectively, dependents expressed their reflections on tourism through temporal objects like hotel ruins, or what I call 'temporal projects' such as a new mega-infrastructural project imagined by the valley's most powerful tourism company (JR) (a cable car, see Chapter 2) or a theatre play (see Chapter 3). These projects present particular cases of dramatization of tourism dependency and its capacity to (un-)make history or the future, but the concerns they reveal have long shaped the everyday of the village, and continue to do so. On the one hand, they highlight how a long history of tourism has fostered a local interest in the past: many locals (typically men over fifty) have written about history, while others have collected historical postcards or participated as lay actors and audience members in historical plays, drawing on the region's vibrant theatre tradition (Bendix 1992, 2018). On the other hand, they also illustrate the inescapabilty of the future as a concern in the valley and beyond. Besides the omnipresent threat of global warming, particularly weighing on the ecosystems and economic structures of Alpine skiing regions (Nöbauer 2022), there was intense debate around JR's new cable car – seen by some as a means of 'securing' the valley's future, while for others it represented a threat to its very existence. In this context, I suggest that, to understand tourism dependency, its uncertainty, its promises and threats, and the ways it shapes issues of belonging and social in-/exclusion, it is essential to look at its temporal orientations.

Such an explicitly temporal approach to tourism has remained rather absent from academic considerations of tourism (for exceptions, see Cohen

1986; Edensor and Holloway 2008; and Crossley 2020). This is particularly surprising given that tourism is shaped by a puzzling temporal paradox: while it is an industry rooted in the critique or escape from capitalist time, it remains deeply embedded within its very logic. This is what anthropologists of tourism Roger Norum and Mary Mostafanezhad describe in one of the few articles addressing time and tourism (Norum and Mostafanezhad 2016). The 'chronopolitics of tourism', they argue, are shaped by an ambiguous relationship between 'capitalist time' and 'authenticity'.[21] In tourism, 'capitalist time' pursues future-oriented ideals of progress, development and growth, as well as nostalgic or escapist imaginaries that break away from capitalist time, often by seeking an 'authentic' connection with what has been destroyed (Leite 2014). Tourism is thus both forward- and backward-looking – an industry that purifies the past and destroys it for the sake of the future.

Who makes these capitalist touristic futures and what they precisely entail varies, especially if different business models – of humanitarian, eco-tourism, slow tourism, community-based tourism – are at hand in different destinations. But, in the dominant model of globalized capitalism, many analysts of tourism agree that this future-making is predatory because it is dictated by logics of growth geared towards expansion and value extraction (Franklin 2008; Mostafanezhad, Azcárate and Norum 2021). In a world shaped by interconnected and competing transnational markets, my work shows that dominant tourism players successfully impose these futures of growth and acceleration as a means to achieve stability, even if they often cause ecological and social damage that local populations lament (see Chapter 2). These favour concerns over the short-term future, where more innovation and acceleration are needed to maintain or 'stabilize' existing achievements in the race of modern capitalism (Rosa, Dörre and Lessenich 2017; López-González and Fennell 2020). As such, the making of tourist futures exemplifies the 'mix of utopic and dystopic representations of time' in capitalism, where time's passage is both creative and destructive of value (Bear 2014b: 78).

In comparison to more recent considerations of the future and tourism development, the realm of the past has been a classical topic of tourism and its study. The Romantics – crucial inspiration figures for the later emerging industry of tourism – formulated critiques of the industrialization and rationalization period in Europe in the eighteenth and nineteenth centuries precisely because they saw these as threatening and alienating for individual human expression (Stone 2014; Watts and Houtman 2024). Thanks to their remoteness, their 'backwardness' (meaning their underdevelopment) and therefore 'unspoilt' character, the Alps became healing sites where the Romantics believed one could reconnect with one's true essence, with oneself, via the experience of the Sublime (Denning 2014). This imaginary of the Alps as authentic was also deeply nostalgic: it associated this region with the past, a remnant of an authentic but disappearing life (Starobinski 1966). In the Alps and elsewhere, the industry of tourism then successfully captured such Romantic critique and affects to attract nostalgic travellers (Graburn 1995; Enzensberger 1996). Based on this genealogy, much of tourism scholars' efforts consisted in explaining how, in the words of one of its most famous theorists, the tourist sought (and failed) to escape the 'alienating', 'workaday' world through 'his' quest of authenticity and wholeness as a 'modern man' (MacCannell 2013 [1976]). Since then, authenticity and nostalgia (often in connection to heritage) have remained central themes in the tourism studies, now approached with greater nuance (Dann 1998; Wang 2000; see also Leite and Graburn 2009).

The nostalgic appeal of the Alps still attracts a lot of tourists, now also in connection with other motivations such as what one could call 'adventure tourism' based on fun, adrenaline and speed. I am here nonetheless primarily interested in a locals' perspective on their own past, in their own valley, interacting with, negotiating and reformulating the tourist's imaginaries of authenticity as has recently been discussed (e.g. Cohen 1988; Salazar and Graburn 2014; de Bernardi 2019). After centuries of touristification, my work speaks of the imposition of an originally externally projected romanticized past on the Alps, and how this manifests ambiguously for tourism dependents. Many problematized it, feeling 'trapped' in the tourist expectations of timelessness and underdevelopment of the Alps (Petite 2013), as well as integrated a romanticized narrative of the past in their perceptions of their own community (see Chapter 3).

While the chronopolitics of tourism shape how tourism is made and shaped as business and consumed by tourists, I focus on how these dynamics also deeply influence the lived reality of tourism dependency. Tourism dependency, I argue here, is shaped by the inherent tension between the conflicting demands of capitalist tourism: creating a future-oriented industry while simultaneously preserving an idealized past. It therefore translates in a backward- *and* forward-looking temporal practice in which inhabitants ponder on tourism's capacity to make history, the present and the future while fearing its upsetting, destructive power over everything that once was.

To capture the temporal nature of tourism dependency, I take my cue from anthropologists of time who have theorized crisis and uncertainty as moments of interrupted normalcy that 'stimulate temporal thought' (Knight and Stewart 2016: 2) and dramatically reveal people's temporal expectations and yearnings (Narotzky and Besnier 2014). Under such circumstances, they have described how pasts and futures make themselves present in the lives of individuals, and how people 'temporally reason' with them (Jansen 2015;

Knight and Stewart 2016; Kleist and Jansen 2016). Such mundane engagements with time often resist and/or contradict the linear succession of events proposed by a normative, Western historicism that orders and separates past, present and future. This scholarship reveals how we perceive time as a result of an intimate and malleable 'social and personal relationship to the past and the future' (Hirsch and Stewart 2005: 262; Moroşanu and Ringel 2016), as well as how it can be dictated by models and ideologies such as nationalism or modernity. My understanding of time lies at the intersection of such existential and structural concerns.

The temporal orientation of the methods adopted during my research also profoundly shapes the presentist perspective I adopt in this book. Conducting ethnographic fieldwork allows for a limited insight on a field from the present of the ethnographer, a present from which she accesses her interlocutors' evocations of the past and future (Hastrup 1990). In line with an undetermined, presentist approach, I consider that 'the past and the future do not exist other than in their representations in the present' (Ringel 2016: 403; Ringel 2018; Bear 2014b). This is not to say that we experience time in a presentist way, as we perpetually 'temporally reason' into the past or the future from the present. But adopting such a presentist lens, in combination with one attuned to people's temporal reasoning, enables the creation of a critical and analytical distance with naturalized temporal narratives such as those of development so present in tourism (see Chapter 2). At the same time, the discussion of temporal projects in chapters 2 and 3 of this book reveal how the efforts and strategies put in place to represent the future and the past have profoundly shaped the present and the everyday of the valley.

Whereas the falling together of different temporal frames is most definitely a feature of human life, Daniel Knight has recently discussed how 'vertigos' are part of the 'affective structure of the Time of Crisis' (Knight 2021: 8). Based on his interlocutors' experiences under conditions of Greek austerity, Knight shows how temporal vertigos manifest in feelings of nausea, dizziness or 'confusion and anxiety about where and when they belong in overarching timelines of pasts and futures' (Knight and Stewart 2016: 32). Temporal vertigos point to the experiences of interruption of normalized linear connection between time frames. This generates an 'anguish of dehistoricization or removal from normalized temporal and historical succession' (Knight 2016: 10) and makes people draw unexpected connections between events they previously saw as distant (Knight 2017). Looking at everyday experiences of people inhabiting a historically, economically and socially stable context like Switzerland might thus be expected to tell us little about their temporal awareness. The story I tell here shows precisely the opposite. It reveals how global capitalism and tourism's intrinsic uncertainty and ambiguity as an industry torn between future-making and preservation of the past manifest in people's temporal preoccupations, their feelings of doubt and fear, in ways that are close to those described in acute crisis contexts (Knight 2015, 2017, 2021). The privileged context of Switzerland and the Alps provides insights into the pervasive and mundane sense of instability that touristic capitalism engenders for those who depend on it, even as they seem to profit from it. As such, my work contributes to a better understanding of the 'doubts' and 'conflicts' of modern capitalism (Bear 2014a) from a place that seems to illustrate its functioning.

This highlights what I have elsewhere described as the 'slipperiness' of tourism dependency (Leitenberg 2023a). If you have ever been in the mountains, you know how quickly a storm can form, how safe trails can suddenly turn into dangerously unstable ground when covered in rain, and how a fall, especially in an inaccessible area, can have tragic consequences. My understanding of slipperiness derives from this experience of the Alpine environment to capture the proximity between promise and threat, success and failure, in the unpredictable but dominant industry of tourism, which relies on the unmanageable nature of other people's wanderlust. It encompasses people's concerns over their capacity to innovate and behave in sync or 'in time' with the accelerating tourism market, as well as their wishes to belong to the rest of the world in a peripheral region, thanks to escapist, touristic imaginaries of timelessness, authenticity and purity to which they also adhere. It also speaks to my local interlocutors' belief in a possible return to times of 'nothingness', poverty and farming without sufficient investment in tourism development, as well as their increasing experience of dissatisfaction and disorientation in seeing the boundaries between past and future eroded in the present.

However, as we shall see, experiencing slipperiness is not always the mark of a negative, submissive state of dependence. It can also mean that you care, or rather that you are made to care, that the developments that characterize tourism speak to you, as a dependent, and that you see yourself in relation to them. Slipperiness can then also be the mark of a certain privilege – something that is particularly striking when contrasted with experiences of entrapment. Such feelings marked the experiences of the migrant hospitality workers I met, restricted in their capacity to 'temporally reason' in the same way as the Swiss and native locals because of being spatio-temporally excluded or 'othered' from the country and the village's life (Fabian 1983; Çağlar 2018). As such, I see slipperiness and entrapment as two uneven and interconnected manifestations of the temporal nature of tourism dependency. Importantly, these are never 'just' temporal, they are also 'thickened' by spatial, embodied, affective, political dimensions (see chapters 4, 5 and 6).

My Alpine field site and its long history of touristification provides an ideal site to explore the slipperiness and entrapment that emerge from structural contradictions between creation and destruction, between the past and the future, which are at the heart of the 'chronopolitics of tourism' and capitalism. This is not to say that these experiences of tourism dependency have not or will not transform. The developments I trace in this book point to the changing dynamics of capitalist tourism in the valley since the end of the twentieth century and the growing power of some over others, as well as the digitalization and acceleration of life and crises (Eriksen 2016). If instability is 'rooted' in capitalism (Harvey 2007), and even more so in touristic capitalism, the forms and ways that this reality manifests are both changing and uncertain.

NATIVISM AND OTHERING IN THE GLOBALIZED ALPS

In resorts visited for their naturalness or authenticity, smallness and marginality often coexist with global connection. The circulation of romanticizing works of art for the past two centuries has led many to think of the Alps as an isolated area, inhabited by naturally rooted and simple or even naive inhabitants (Kilani 1984; Schär 2012). Contrasting with such stereotypes, many anthropologists working on the region have shown how the Alps is a complex border region, shaped over centuries by mobility patterns and interconnections with the rest of the world (e.g. Viazzo 1989; Membretti and Viazzo 2017; Cretton, Boscoboinik and Friedli 2020; Nöbauer 2022; Burke et al. 2023; Schneider and Tauber 2024). The lack of opportunities that its inhabitants may experience – and many do – is the result of a process of marginalization that is subject to action and change. At the same time, this remoteness is also a 'resource' or a capital (Saxer and Andersson 2019) that has attracted new inhabitants in search of a greater connection with nature and a better work–life balance (Boscoboinik and Cretton 2017). These authors' descriptions of mobility and interconnection as fundamental features of the Alps, even in non-touristic regions, are particularly relevant for understanding the power of the developmentalist tourism narrative. If tourism is not the only means for Alpine communities to achieve mobility and connectivity, it has come to be perceived as such.

Without subscribing to the simplistic narratives of backward, isolated mountain regions, this book engages with the work of those who have portrayed the Alps as spaces of critique towards the urbanized world – not only in the eyes of artists and intellectuals from the lowlands, but also the inhabitants of the Alps themselves (Stacul 2003a, b; Stacul 2006; Lems 2023). In their seminal work *The Hidden Frontier* on ethnicity and ecology in the Italian Alps, Cole and Wolf have shown that this critical stance is particularly important in the German-speaking context. Comparing the

Italian-speaking village of Tret and that of German-speaking St Felix, they reveal how dwellers of the latter associated their rural world with autonomy and a superior, healthier world, as opposed to the former, who viewed the city as 'the seat of civilization and urbanity' (Cole and Wolf 1999: 10). For German-speaking Alpine dwellers especially, the idea of the *Heimat*, which stands for homeland but also the more material definition of home and farm, represents the centre of one's material and cultural world (Applegate 1990; Leimgruber 2019).

The Alpine villages analysed by Wolf and Cole in the 1970s surely present contexts that are different from the one I examine here: my field site is now a relatively wealthy village, and a touristic resort visited by countless visitors who largely outweigh the number of locals. Yet, although tourism may well be the cosmopolitan activity par excellence (Salazar and Graburn 2014), my findings reveal how this industry has done more than simply replace the forms of rooted belonging in the valley that could be associated with rurality and farming. Instead, ideals of autochthony and rootedness informed the attitudes of many locals, who insisted on their authentic belonging as 'original' inhabitants of the village, the descendants of poor farmers, who were truly 'at home' in it, because they had been born and raised in the village, carried local last names, spoke typical local dialect, had a long family history in the municipality and/or had it inscribed as their 'place of origin' on their identity cards.[22] Some institutions were also particularly emblematic of this native identity, such as the common property institutions like the Alpine corporations, present in many mountain regions (Haller et al. 2019). Alpine farming, though largely supplanted by the tourism industry, had endured as a symbol of nativeness, preserving its cultural and traditional essence.

As such, this ethnography draws a parallel with what is the last ethnographic monography on the Swiss Alps in English, written by the American anthropologist Robert Netting, who explored the role of pastoral commons in the village of Törbel (Netting 1981).[23] The 'ecological balance' sought by his interlocutors was also a concern in my field site, where the increasing pressure imposed by tourism and capitalist actors on these resources and institutions was deeply felt (see Chapter 2). However, where farming marked the subsistence of Netting's interlocutors, in the Bernese resort where I conducted research it was a distant but reactivated reality that participated in defining genuine native belonging to the valley. This situation raises the question of how, and through what social efforts, the exclusive category of 'natives' could be evoked and recreated within a complex world. Exploring the role that the farming past played in shaping these evocations, I adopt a perspective that social scientists have described as 'boundary-work' (Barth 1998; Lamont and Molnár 2002) on belonging and exclusion. This implies an attention paid to the processes, the social efforts, or 'work' through which many of my *Einheimische* interlocutors (among others)[24] actively performed these identities in opposition to tourists, outsiders, foreigners, lowlanders, and such like.

To be clear, this is not a study of an Alpine village's 'native' population; nor do I wish to imply that the natives I met neatly fit into this category. Like any other population, the natives I met were characterized by their heterogeneity. Nevertheless, many in the valley emphasized the profound distinctiveness – or even incommensurability – of native identities in a world marked by increasing acceleration and interconnectedness. These entailed a conservative, down-to-earth, healthier lifestyle typical of Alpine 'village life' that many locals, as well as (Western) tourists, believed tended to disappear in the urbanized and more progressive lowlands. Thus, many of my interlocutors explained that it was only natural that the natives should be given primacy and priority, as original inhabitants, when it came to finding accommodation or jobs in the region or when expressing opinions on the valley's development. In other words, they subscribed to De Genova's definition of nativism[25] as a 'preference for the "native" exclusively on the grounds of "*being* native"' (De Genova 2016: 233, italics in original; see Chapter 1).

Nativism constituted the dominant 'politics of belonging' at play in the village, to refer to a concept coined by sociologist Nira Yuval-Davis (2006). This entailed deeply intimate feelings, in this case 'homely' feelings linked to ideals of *Heimatverbundenheit* (attachment to the homeland), as well as highly political dimensions, creating boundaries between those who belong and those who do not, based on the more or less institutionalized identity markers mentioned above. These local attitudes were also embedded in nativist national politics, portraying Alpine dwellers as the embodiment of the 'authentic' Swiss identity, with mythical roots deeply tied to the Alps (Schär 2012). They speak of a local and national context shaped by the successes of the nativist, right-wing conservative Swiss People's Party (SVP), as well. The Alpine nativism I encountered in the valley sometimes took the shape of localism, sometimes national nativism, and often both.

Throughout the following chapters, the Alpine village emerges as a paradoxical space – both deeply embedded in globalized networks and marked by conservative, nativist tendencies. These tensions – between tradition and change, isolation and connection, conservation and development – are deeply connected to the temporal paradoxes that define tourism dependency, as a state marked by the opposing demands of preserving the past and advancing the future. They also translate in attitudes – of embrace or rejection – between self and other, be it non-native locals, migrants, 'mass tourists' and returning, historical guests (*Stammgäste*). These tensions serve as the backbone of the story told here, offering a lens through which to examine how various actors negotiate and respond to their competing dynamics. At the same time, the situation of this Swiss Alpine village reflects broader global realities, increasingly shaped by the interplay of nativist movements and global interconnection (see Guia 2016).

MIGRANT HOSPITALITY LABOUR

In the valley, the spectrum of non-indigeneity stretched from *Auswärtige*, 'outsiders' (literally from another place), who were other Swiss locals who did not draw on markers of nativeness – such as having been born in the valley, or having typical names or dialect, but spoke Swiss-German and were relatively well-included in the local community – to *Ausländer*, foreigners or migrants, who represented the majority of the valley's workforce in hotels and restaurants. The latter belonged to the 'toured ones' – they were dependent on tourism too – but their belonging to the village was contested by many native or Swiss locals, and sometimes also by migrants themselves. This position, its historical roots and social consequences, highlights another critical aspect of tourism dependency: its unevenness.

In defining the issue of migrant tourism labour in this way, I do not suggest that the distinction between tourism and migration is clear-cut. Extensive scholarship has illuminated the blurred boundaries between the two, emphasizing the risks of adopting a Western-centric perspective that may obscure their intricate and intersecting dynamics (Adams 2022; Bloch and Adams 2022; Vietti 2024). However, my perspective on migrant labour is again shaped by the emic insights of migrant locals, who expressed a keen and deeply felt awareness that their lives were fundamentally different from those of the tourists they were often tasked with serving. As one migrant worker I encountered aptly put it, tourism labour for her was not merely about 'living *off* tourism but also 'living *for* it'. It is this all-encompassing presence of tourism labour in one's life that I seek to explore and interrogate here.

Migrant presence in the valley, the Bernese Highlands and the Alps draws on a long history. By the end of the nineteenth century, workers from Northern Italy were migrating to the Jungfrau Region to participate in the country's rapid industrialization and construction of a dense railway system, often at the risk of losing their lives (Moser 1997). At the same time, women from other regions of Switzerland and migrants from Italy, Germany and Austria were being employed in hotels and restaurants. Throughout the twentieth century, the local, like the national, hospitality sector developed thanks to a migrant workforce from Italy, Spain, Yugoslavia and Portugal (see Graf 2021). As of 2017, migrants represented 24 per cent of the village population.[26]

The village's development, driven by a migrant workforce, exemplifies a global trend of dependence on minority or marginalized populations to perform essential labour under precarious conditions (Seiger et al. 2020; Bastos, Novoa and Salazar 2021). Those who cross borders are stripped off their qualification, experiences and opportunities, and pushed to perform jobs that often endanger their long-term financial and existential well-being (Barber and Lem 2018). In the tourism industry and the hospitality sector, a large number of social scientists have discussed how racialized and ethnic minorities, women and migrants have long represented the majority of the workforce (Janta et al. 2011; Ladkin 2014; Ioannides and Zampoukos 2018; Janta and Ladkin 2024).[27] They have described how these populations have been integrated in a labour-intensive business in perpetual search of unqualified labour, to perform tasks marked by a difficult working environment with irregular working hours, short-term contracts based on seasonality, low salaries, physical and emotional stress, and the general absence of unions (Baum 2007). In doing so, these scholars have revealed the importance of considering tourism development together with the effects of migration regimes that are strongly differentiating workers according to their provenance (McDowell, Batnitzky and Dyer 2007, 2009).

In one of the few ethnographic monographies published on the topic of hospitality labour in touristic resorts, Adler and Adler (2004) describe the strong hierarchies differentiating types of workers in Hawaii: the lower, consisting of immigrants from developing countries (the 'newcomers') and 'brown-skinned' locals of 'Asian, Polynesian, Pacific Island, Hawaiian, and Portuguese descent' ('the locals'); and the higher, made up of young, educated, middle-class individuals from mainland America seeking an alternative lifestyle (the 'seekers') and career-oriented 'managers' from mainland American involved in top-level management. Their analysis highlights the ethnic stratification within hotel resorts, while also drawing a critical distinction between migrants and locals on one side, and seekers and managers on the other. The first, significantly larger in number, is characterized as 'trapped labourers' being 'completely subject to the economic vicissitudes of the local labour market' (ibid.: 6). By contrast, the second group is described as 'transient' due to the greater flexibility they possess in moving in and out of the resort.

In the Swiss Alps, like in Hawaii, the crowd of hospitality workers I encountered was made up of different people, of various statuses and provenance. Individuals with the lowest levels of language proficiency and education, often hailing from Southern or Eastern Europe, were typically assigned to roles commonly referred to in tourism literature as the 'backstages' of tourism labour (Urry and Larsen 2011; Kensbock et al. 2016). These jobs were characterized by their invisibility to the tourist, working spaces (kitchens, washing rooms) inaccessible to them, or during times when they were absent (cleaning rooms).[28] Those who could speak German were moved to 'front-stage jobs' interacting with customers, and those who spoke German as a mother tongue often occupied managerial positions, overseeing the other groups of workers. As many hospitality workers and hoteliers themselves told me, the Swiss made for a small minority among workers, usually as managers. Native locals were almost completely absent in the hospitality sector, except for Adrian, whom I introduced at the outset of this Introduction and who insisted on his exceptional position as native local *and* hotel worker. His fellow native locals avoided this sector not only due to its poor working conditions but also because of the sociocultural meanings attached to hospitality work (see Chapter 5). This stood in stark contrast to the affinity many locals displayed for other forms of tourism labour, such as working on ski lifts or JR's transportation infrastructure.

While united in their dependency on tourism, both vulnerable or 'subject to the vicissitudes' (Adler and Adler 2004: 6) of the tourism market, the migrants and native locals in the valley experienced this state very differently.[29] Many of the hospitality migrants I spoke with were determined to remain in the valley and establish their futures within the village. However, they faced significant obstacles, as the anticipated opportunities for social and existential mobility (Hage 2009) promised by tourism and migrant labour failed to materialize – often with quite the opposite effect. Cast as outsiders, but contributing to the industry from the inside, they also spoke of the highly politicized process of integrating into the conservative Alps but also Switzerland, a country with some of the most restrictive citizenship regulations in Europe (Schmid 2023). Continuously evaluated in terms of deficit, with the looming threat of deportation, they found themselves at the bottom of the local and national politics of belonging, and turned to the hospitality industry as an entry into the jobs market of wealthy Switzerland from below, overtaking activities that were largely disregarded by citizens (see also Graf 2021). In spite of having spent years in the resort contributing to its tourist economy, and although a historical reality, my work among migrants, following their often barely visible traces in the valley's history of successful touristification, was marked by a sense of exclusion and temporariness.

The case of the touristic Swiss Alps, likely similar to other European resorts (e.g. Marcher et al. 2022), prompts us to rethink the dynamics revealed by Adler and Adler in Hawaii's luxury hotels (2004). Contrary to what they demonstrate of the 'brown skinned' locals' position, local and national belonging in European contexts is often intertwined with privilege and therefore marked by other experiences of tourism dependency. This does not mean that these experiences are not marked by difficulties – on the contrary, in the Alps, processes of domination, of dispossession and increasing instability or precarity shape the conditions within which all inhabitants find themselves. This reflects tourism's 'predatory' logics (Azcárate 2020) but also its global backdrop of uncertainty, defined by a context of acceleration of crises (Eriksen 2016) and of ecological, social and financial precarity (Tsing 2015).

Acknowledging this situation, mobility and migration scholars Ayse Çağlar and Nina Glick Schiller (Çağlar and Glick Schiller 2018; Glick Schiller 2021) have called to move away from the sole focus on migrants in refugee, migration and mobility studies.[30] They have argued for the examination of both migrants and non-migrants under the same analytical lens and have pleaded for a study of the general processes of dispossession that (unevenly) impact populations across migratory statuses (see also Ramsay 2019). Inspired by this approach, this book seeks to illuminate the shared realities of tourism dependency, focusing on both those who migrate (mobile hospitality workers) and those who stay put (Swiss and/or native locals) in a context marked by uncertainty, while highlighting its uneven impacts on populations of varying statuses and categories. In doing so, it draws attention to relationships that cannot be neatly compartmentalized into separate areas of scholarship, such as mobility, migration, tourism, capitalism, or Alpine studies.

VILLAGE ETHNOGRAPHY IN A CONNECTED WORLD

The work presented here is based on data gathered during the majority of my fieldwork, which took place between September 2017 and August 2018, followed by shorter stays and visits between January 2019 and May 2023. The vastness of the valley's territory, it being one of the biggest municipalities in Switzerland, combined with the frenetic rhythm of the tourism economy, soon became serious challenges for my research. With everyone running around me, between activities, jobs, guests and tasks, I struggled to establish relationships beyond the very limited time that tourism encounters offered – a common difficulty when researching in the fleeting, complex and uncertain economy of tourism (Graburn 2002; Kaaristo 2018; Leite, Castañeda and Adams 2019). My decision to focus on tourism dependency – a reality experienced by those who, like me, lived in the village in the everyday – responded to this challenge.

To get a sense of the complexity and reality of this particular place and those who lived in it, I turned to one of the foundational tools of anthropological research: village ethnography. Many have unveiled the problematic assumptions of village ethnography and its 'rooting' of extra-European community of natives in places and cultures that could be 'discovered' by the ethnographer (Malkki 1997; Gupta and Ferguson 1992; Amit 2000). In a world shaped by globalization and in a transnational context like the tourist resort I conducted research in, this choice might seem surprising, if not anachronistic. Conducting a village ethnography in the Swiss Alps, however, proved to be a fundamental tool for exploring how humans navigate some of the greatest challenges of our time, be they environmental or social. As the insightful critiques of Matei Candea (2007) and Michael Herzfeld (2015) have revealed, village ethnography needs to be recalibrated for the study of the contemporary, instead of relegated to the background. Such a task involves complicating what we understand of 'the village' and its inhabitants, and how we mobilize these notions to address particular questions. Village ethnography, as I understand it here, thus comprises both openness and boundedness (Herzfeld 2015: 338). Rather than contradictions, such features become objects of study, to reveal the ongoing significance given to place and local, regional or national particularism in a globalized world for both the industry of tourism and political nativism.

Conducting village ethnography as a non-holistic, bounded and open enterprise does not involve talking with everyone or giving a full overview of what happens in the village. Rather, it interrogates how locality and sitedness transpire in a transnational context, and how they are experienced by a selection of actors. As mentioned earlier, this village ethnography is thus centred on the particular categories of people I followed in their everyday experiences because of what they exemplified of the village dwellers' unequal dependency on tourism: the so-called insiders (the Swiss native locals) and the outsiders (migrant hospitality workers). This is not to say that I wish to challenge the fluidity of such categories or the actors situated in between, such as the non-native Swiss locals. Like the tourists, they come and go in this study, but they are not its main protagonists.

To understand how the local financial, social and moral economy of tourism worked and to approach the reality of tourism dependency, I conducted participant observation in different private settings. As a social anthropologist, I could attend private general assemblies of regional and local businesses, tourism lobbyist meetings, club meetings, project launches, political meetings and municipal assemblies. I was also authorized to work in a hotel for four half-days in order to conduct participant observation among the hospitality workers. Next to these more formal settings, I spent time with village inhabitants (both 'insiders' and 'outsiders' and those who were neither *Einheimische* nor migrants) and their families, accompanied them on short trips, stayed in their houses, helped them in everyday chores, and regularly visited them at work. I also conducted participant observation in multiple tourist settings, stayed in hotels as a guest, went on numerous guided tours, attended countless organized tourist events, and used the local touristic infrastructures on a daily basis. Next to the many informal discussions I had on these occasions and during my quotidian chores, and during my day-to-day life in the village, I conducted around seventy-five recorded interviews and a handful of written ones. Approximately half of these were expert-interviews with tourism lobbyists, hotel owners, political leaders, association members and so on, the other half being with inhabitants and biographically oriented. I conducted all interviews in German, English or French, although I relied on assistance for some of the transcriptions of interviews in local dialect. All English translations are mine, and I cite the original statements in standard German and not in Swiss-German (if not mentioned otherwise), for reasons of feasibility and consistency. While I usually met with such experts once, I met at least twice (usually three times) with a dozen informants for biographical interviews that lasted from one to three hours. I also relied heavily on the communal archives, which I visited in order to access information on the population, as well as newspaper articles, reports and the like.

Reflecting the variety of profiles in a village, my research involved studying 'up', 'down' and in-between, with interlocutors who had more and less power than me, making use of my insider–outsider position, as a Swiss and as someone from a different cultural and linguistic background than that of the locals (Mughal 2015; Leitenberg 2023b). I talked to local elites, such as tourism lobbyists, who mobilized particular knowledge, expertise and power I did not have, and to natives whose belonging was naturalized, but also to denizens whose precarious living conditions were not comparable to mine, as a Swiss citizen. With others, such as Swiss locals who had immigrated to the village for work from urban centres, I was on a more equal footing. In all these situations, my positionality as a French-speaking Swiss female anthropologist from the 'lowlands' shaped my relationships and my research.

This position also had geographical implications, as I lived in the centre of the village for most of my research, where most of the touristic events took place, surrounded by guests and dwellers from different horizons, who were often more urban-minded. Finding an accommodation online proved to be difficult and expensive, and it was only after visits in the village, looking for ads in the local shops, and then insisting that I was staying longer than the season – and was therefore working there, in the eyes of locals – that I was able to find something affordable. Contrary to my expectations, living in the centre of the village did not allow me to form strong and immediate relationships, as rural village life often implies (Belina 2021) – at least not with everyone. The more I seemed to become familiar with those who were described as the 'insiders' of the village, the farther I moved away from the village's *Hauptstrasse*. My stays were punctuated by rather long trips, usually going up the valley's flanks by bus to talk to those who lived on the outskirts of the village, in the (remnants of the) *Streusiedlung*, but were in fact its 'natives'. Migrant hospitality workers and other outsiders, by contrast, lived in more direct proximity to me, in furnished studios or rented apartments. Studying 'up' and 'down' was not just a matter of power, as Nader (1972) famously implied, but also had topographical connotations that had central implications for people's way of engaging with tourists, as I will develop throughout the next chapters.

As a Swiss anthropologist studying the Alps as the epitome of Swissness (Zimmer 1998), it was sometimes assumed that I would share a degree of familiarity or at least cultural affinity with my field site because of my citizenship. This was often contradicted by my local interlocutors, who regularly insisted on my 'outsider' background, sometimes even questioning my Swissness. This relied on the strong localist discourses discriminating between 'genuine' natives (*Einheimische*) and Swiss outsiders (*Auswärtige*), as well as broader national narratives on the cultural divide or *Röstigraben* (the rösti ditch) between the Swiss-German and the Swiss-French parts (Wimmer 2011). For those excluded from the local but also the national community, such as the migrant hospitality workers, I was very much an insider. Unlike them, I did not rely on a precarious working contract, my presence in the valley was not tied to revokable residency permits, and I benefited from a comfortable national grant and a stable financial situation. These workers from Portugal, Germany, France or Italy regularly asked me – often as the only Swiss they knew outside of work – for help, information and advice regarding residency permits, citizenship tests, administrative tasks and school matters for their children.

ABOUT THIS BOOK

Although they speak of the structural realities created by capitalism and tourism, the questions that animate this book cannot be approached without the deeply intimate experiences of people who depend on tourism to make ends meet. To counterbalance the fleeting nature of the village as a 'touristic borderzone' (Bruner 2005), each of the chapters follows a person whose experience casts a different light on tourism reliance and its temporal manifestations in the Alps. These characters' positions and views on tourism all display the heterogeneous and fluctuating nature of dependency. They appear in their capacity to navigate tourism's ambiguity instead of passively living off it: each chapter reveals how people reflect on, mobilize, address and (attempt to) modify tourism's development more or less successfully, in ways that resist fixed boundaries between dependent and independent.

The first chapter takes a detailed look at the historical and political context of the valley, as a touristic resort and a place shaped by nativism and anticosmopolitanism (Chapter 1). The chronopolitics of tourism feature in the second and third chapters, based on the examination of two temporal projects – the new cable car and a local amateur theatre play – that both reveal a future-oriented, capitalist industry based on linear growth and acceleration into the future (Chapter 2), as well as nostalgic imaginaries that serve to attract guests but also feed into local discourses of uniqueness (Chapter 3). These projects problematize the temporal demands of tourism and are emblematic of how tourism domesticates and unsettles, as well as celebrates, the natural and human environment. In Chapter 4, I look at how these tensions were experienced and navigated for locals who felt torn between the race forward imposed by the tourism industry and a yearning for the authentic past, also informed by nostalgic, Romantic imaginaries. Whereas this points to intrinsic paradoxes of tourism, this chapter reveals the recent developments of this tension.

In chapters 5 and 6, I bring back the deeply politicized matter of belonging and difference into the equation of tourism dependency. While all forms of dependency are unique, this part sheds light on the structural dependency of the local, Swiss and transnational economy on migrant labour for a functioning economy of tourism. Migrants themselves also depended on tourism labour in Switzerland as a vehicle for a better future and a more stable life. Chapter 5 looks at this situation of interdependency, often described as 'win–win', to then shed light on how, for more than 150 years, migrant workers have often been othered and made invisible in Switzerland and specifically in the village. The temporal projects (chapters 2 and 3) that were said to gather and/or divide the village dwellers appear as the exclusive concern of only some of them.

Chapter 6 weighs up the moral paradoxes of tourism as a promise and a threat to explore how unequally and unevenly such contradictions manifest themselves in the lives of migrant hospitality workers. In the Conclusion, I address the recent developments of tourism dependency in the aftermath of the Covid-19 pandemic, the relationships between uneven forms of tourism dependency and their manifestation in people's environments, lives and bodies, as well as the imbrication between capitalism and nativism.

NOTES

1. This includes all people with permanent residency and approximately 300 people with 'weekly residency' (*Wochenaufenthalter*) in the village, meaning with their main residency located elsewhere.

2. This also recalls tourism's importance globally, since its democratization in the 1970s, making up 10.3 per cent of global gross domestic product (GDP) in 2019: https://wttc.org/research/economic-impact (last accessed 16 June 2023). I am using data from 2019 as the global tourism industry is still recovering from the impacts of the Covid-19 pandemic at the time of the redaction.
3. These were, for instance, mercenaries rented out to European powers in the seventeenth century (Netting 1981), or dwellers who migrated to North America during the eighteenth and nineteenth centuries to find land and resources, and gain power (Schelbert 1976). See Chapter 1.
4. Along with its already existing difficulties, mountain farming had suffered from a market liberalization at the end of the twentieth century that favoured the emergence of bigger farms from the flatlands and caused mountain farming to become increasingly less productive (Tiefenbach 2013).
5. British author Leslie Stephen even characterized the Swiss Alps as 'the Playground of Europe' (Stephen 1895).
6. This, in many ways, reflects the omnipresence and fuzziness of tourism; at once a process, a social field, a cultural and historical phenomenon, a socioeconomic exchange and an individual and collective experience, tourism engages a multiplicity of practices and a wide array of actors and institutions (Leite and Graburn 2009).
7. https://www.bfs.admin.ch/bfs/de/home/statistiken/regionalstatistik/regionale-portraets-kennzahlen/gemeinden.assetdetail.11587763.html (last accessed 9 June 2023)
8. Although the tendency in the Bernese Highlands tends to improve, it remains a fragile region where outmigration constitutes a threat, like in many Alpine valleys. See https://www.bernerzeitung.ch/die-berner-oberlaender-bevoelkerung-unter-der-lupe-284763144085andhttps://archive.letemps.ch/archive/www.letemps.ch/grand-format/population-immigration-vieillissement-decouvrez-levolution-suisse-15-graphiques.html (both last accessed 28 March 2023).
9. In the 1980s, researchers concluded that the region was characterized by its 'extreme monostructure' based on tourism (Messerli 1987). They nonetheless also noted the symbiotic relationship between mountain farming and tourism, as the farmers relied on tourism and vice versa (Wiesmann and Liechti 2004).
10. Only one much smaller business remained, owned by a rival of JR, consisting of a small gondola, with a restaurant, a summer sledge and, more recently, a 'fly-line', open from May until October. However, the influence of this business is incomparable to JR, which parent company (the Jungfraubahn Holding AG or Jungfrau Railway Group) owns infrastructure across the entire Jungfrau region.
11. In 2019, tourism represented a quarter of new jobs created across the world, before the setbacks following the outbreak of Covid-19. See: https://wttc.org/Portals/0/Documents/Reports/2021/Global%20Economic%20Impact%20and%20Trends%202021.pdf (last accessed 16 June 2023).
12. This recalls Besky's description of monoculture in the colonial context of Indian tea plantations: 'monoculture sutures people's identities to single things' (2017).
13. Such as presented by Graburn 1977; Cohen 1979; Smith 1989; MacCannell 2013.
14. This does not mean that all foreigners in the village were working in the hotels or restaurants. However, given the important reliance on this type of workforce, present in all restaurants and hotels, it probably accounted for the majority. See note 27.

15. See: https://www.imf.org/en/Publications/fandd/issues/2020/12/impact-of-the-pandemic-on-tourism-behsudi (last accessed 8 June 2024). Since the outbreak of Covid-19, debates on the future of tourism, for instance, took place at the European Council in 2021 and 2022, revealing the interest of public and private actors in the matter. https://www.consilium.europa.eu/en/press/press-releases/2021/05/27/tourism-in-europe-for-the-next-decade-council-adopts-conclusions/, and: https://www.unwto.org/news/unwto-and-european-commission-share-joint-vision-for-tourisms-future (last accessed 20 Sept. 2023).
16. Broadly speaking, the dependency theory explains global inequalities by emphasizing the exploitative relationships between developed (core) countries and developing (periphery) countries. It explains how the 'underdevelopment' in peripheral nations is not a result of internal deficiencies but a consequence of their historical and structural dependence on more industrialized nations (Bianchi 2018).
17. Various authors have also discussed how assumptions about the positive or negative qualities of tourism often simplify and essentialize our conceptions of the community, of who 'counts' as a local or a host (Tucker 2003; Stronza 2005; Kaaristo 2018).
18. Bianchi adds: 'This view was reinforced by the fact that attempts by "Third World" states to pursue strategies of economic self-reliance through state-led tourism development, for example in Grenada, Jamaica, Tanzania, had met with limited success' (Bianchi 2018: 91).
19. The context in which Ferguson (after Bayart 2000) proposed to understand this dependency as a 'mode of action' is different from that of my field site in that the latter focused on the strategies put in place by poorer populations in Africa in order to 'draw on the income and wealth of better-resourced others' (Ferguson 2015: 107). Ferguson explains that dependency plays a central role in the 'distributive labour' put in place by these populations, actively embedding themselves in various relationships of dependence (formal or informal) to access resources. The financial stability enjoyed by the majority of my informants shifted the stakes of dependency in the Swiss valley: they concerned the ability to critically reflect on and modify an industry of tourism that *already* provides access to resources. The isolation and precarity experienced by migrant workers I present in chapters 5 and 6 could however be read as serious challenges posed to this capacity to perform 'distribution labour' via relations of dependency.
20. Regarding the potentiality to be 'nothing' if it was not for tourism, Azcárate explains: 'Often, particularly among indigenous peoples working at beach resorts or luxury hacienda hotels, [tourism] was used to conjure up a centuries-long, thriving monocrop agro-industrial landscape that had turned to ruins, leaving people devoid of economic alternatives by the 1970s, but which people 'had grown used to'. Being stuck with tourism, then, also established a comparison with the way things were in the past. . . . As an expression, 'stuck with tourism' reminded people that state-led tourism was a long-term response to another economic and social crisis' (Azcárate 2020: 17).
21. Norum and Mostafanazhed identify a third aspect of the 'chronopolitics of tourism' in 'ecology'. As, in the tradition of Romanticism, nostalgia merges imaginaries of authenticity of places and understanding of the Alps as a 'pure' environment, I understand the authenticity and the ecology as part of the same realm (see Chapter 3).

22. Place of origin translates into *Heimatort* in German and is a legal category for every Swiss citizen; it defines municipal citizenship, where their ancestors lived, and where their family name comes from. The *Heimatort* is inherited through the father, according to the principle of *jus soli*.
23. And many other works that have, mainly, focused on the changing relationship between farming and the natural/social environment (Jaeggi 1965; Krauß 2018), often in the Swiss Alpine region of Valais (Niederer, Bätzing and Anderegg 1993; Berthoud 2001; see Leimgruber 2005 for a review).
24. As the next chapters reveal, they were joined by other Swiss locals and tourists.
25. My approach here differs from recent discussions of 'neo-nativist' movements in Europe (Guia 2016; Betz 2017). These have also grown in Switzerland over the last years, for instance in reaction to immigration or the state's governance of the Covid-19 pandemic. My findings nonetheless point to a longer tradition of nativism that I will describe in Chapter 1.
26. This was also indicated in the local booklet handed out to all new inhabitants.
27. The Swiss hospitality sector drew on 50 per cent of migrant workers as of 2021, a number increasing in lower-tier positions (Nathani et al. 2022: 10–11).
28. I did not observe nor could I collect data on racialized workers occupying the least visible positions in the industry, as is often reported in the literature on hospitality work (Adler and Adler 2004; Azcárate 2020). However, my access to the backstages of hotels was limited (I could only visit three). Worker Rachid, who I present in Chapter 6, was the only racialized worker I met, and he performed jobs with great degrees of interaction because of his language proficiency. Nevertheless, the racial discrimination in hospitality businesses has been described in great detail in other Swiss contexts (Kalbermatter 2020).
29. This contrast between hospitality migrants and native locals, and what it reveals of an acute but differently experienced tourism dependency, led me to focus primarily on them instead of the 'seekers' or lifestyle migrants, who were also present in the village but had a more marginal position. Because of their education, age and provenance, these workers experienced a tourism dependency that was not 'problematized' in the same way it was for the two other groups; however, Chapter 6 will reveal how 'seeker' positions may also turn into entrapping hospitality migrant positions.
30. Janine Dahinden (2016) has similarly called to 'demigranticize' mobility and migration research, but more as a way to move beyond the essentializing categories of the nation-state or the migration apparatus.

REFERENCES

Adams, Kathleen M. 2022. 'What Western Tourism Concepts Obscure: Intersections of Migration and Tourism in Indonesia', in Harng Luh Sin, Mary Mostafanezhad and Joseph M. Cheer (eds), *Recentering Tourism Geographies in the 'Asian Century'*. Abingdon: Routledge, pp. 30–55.

Adams, Vincanne. 1992. 'Tourism and Sherpas, Nepal: Reconstruction of Reciprocity', *Annals of Tourism Research* 19(3): 534–54.

Adler, Patricia A., and Peter Adler. 2004. *Paradise Laborers: Hotel Work in the Global Economy*. New York: Cornell University Press.

Amit, Vered. 2000. *Constructing the Field: Ethnographic Fieldwork in the Contemporary World*. London [u.a.]: Routledge.

Applegate, Celia. 1990. *A Nation of Provincials: The German Idea of Heimat*. Berkeley: University of California Press.

Ars, Mojca Stubelj, and Marko Bohanec. 2010. 'Towards the Ecotourism: A Decision Support Model for the Assessment of Sustainability of Mountain Huts in the Alps', *Journal of Environmental Management* 91(12): 2554–64.

Azcárate, Matilde Córdoba. 2020. *Stuck with Tourism: Space, Power, and Labor in Contemporary Yucatan*. Oakland: University of California Press.

Barber, Pauline Gardiner, and Winnie Lem (eds). 2018. *Migration, Temporality, and Capitalism: Entangled Mobilities across Global Spaces*. Cham, Switzerland: Springer.

Barth, Fredrik (ed.). 1998. *Ethnic Groups and Boundaries: The Social Organization of Culture Difference*. Long Grove, IL: Waveland Press.

Barton, Susan. 2008. *Healthy Living in the Alps: The Origins of Winter Tourism in Switzerland, 1860–1914*. Manchester: Manchester University Press.

Bastin, R. 1984. 'Small Island Tourism: Development or Dependency?', *Development Policy Review* 2: 79–90. https://doi.org/10.1111/j.1467-7679.1984.tb00491.

Bastos, Cristiana, Andre Novoa and Noel B. Salazar. 2021. 'Mobile Labour: An Introduction', *Mobilities* 16(2): 155–63. https://doi.org/10.1080/17450101.2021.1885840.

Bätzing, Werner. 2015. *Die Alpen: Geschichte und Zukunft einer europäischen Kulturlandschaft*. Munich: C.H. Beck.

Baum, Tom. 2007. 'Human Resources in Tourism: Still Waiting for Change', *Tourism Management* 28(6): 1383–99.

Bayart, Jean-François. 2000. 'Africa in the World: A History of Extraversion', *African Affairs* 99(395): 217–67.

Bear, Laura. 2014a. 'Doubt, Conflict, Mediation: The Anthropology of Modern Time', *Journal of the Royal Anthropological Institute* 20(S1): 3–30.

———. 2014b. 'For Labour: Ajeet's Accident and the Ethics of Technological Fixes in Time', *Journal of the Royal Anthropological Institute* 20(S1): 71–88.

Belina, Bernd. 2021. '"Provinzialität" bei Adorno', *Geographische Zeitschrift* 109(2–3): 105–25.

Bendix, Regina. 1992. 'National Sentiment in the Enactment and Discourse of Swiss Political Ritual', *American Ethnologist* 19(4): 768–90.

———. 2018. *Culture and Value: Tourism, Heritage, and Property*. Bloomington: Indiana University Press.

Berthoud, Gérald. 2001. 'The "Spirit of the Alps" and the Making of Political and Economic Modernity in Switzerland', *Social Anthropology* 9(1): 81–94.

Berwert, Adrian, and Antoine de Bary. 2005. *Empirische Ermittlung von Tourismusanteilen einzelner Wirtschaftszweige: Tourismusgemeinden der Region Berner Oberland*. Rüschlikon: Rütter and Partner.

Besky, Sarah. 2017. 'Monoculture', *Cultural Anthropology Website* (Theorizing the Contemporary).

Betz, Hans-Georg. 2017. 'Nativism across Time and Space', *Swiss Political Science Review* 23(4): 335–53. https://doi.org/10.1111/spsr.12260.

Bianchi, Raoul. 2000. 'Migrant Tourist-Workers: Exploring the "Contact Zones" of Post-Industrial Tourism', *Current Issues in Tourism* 3(2): 107–37.

———. 2018. 'The Political Economy of Tourism Development: A Critical Review', *Annals of Tourism Research* 70: 88–102.

Bloch, Natalia, and Kathleen M. Adams (eds). 2022. *Intersections of Tourism, Migration, and Exile*. London: Routledge.

Boissevain, Jeremy. 1996. *Coping with Tourists: European Reactions to Mass Tourism*, Vol. 1. Oxford: Berghahn Books.

Boscoboinik, A., and V. Cretton. 2017. '"Find Your Nature" in the Swiss Alps: In Search of a Better Life in the Mountains', *Český lid* 104: 199–212.

Breiby, Monica A., Hogne Øian and Øystein Aas. 2021. 'Good, Bad or Ugly Tourism? Sustainability Discourses in Nature-based Tourism', in Peter Fredman and Jan Vidar Haukeland (eds), *Nordic Perspectives on Nature-based Tourism: From Place-based Resources to Value-added Experiences*, 130–43. Cheltenham: Edward Elgar Publishing.

Britton, Stephen G. 1980. 'The Evolution of a Colonial Space-Economy: The Case of Fiji', *Journal of Historical Geography* 6(3): 251–74.

———. 1982. 'The Political Economy of Tourism in the Third World', *Annals of Tourism Research* 9(3): 331–58.

Bruner, Edward M. 1996. 'Tourism in the Balinese Borderzone', in Smadar Lavie and Ted Swedenburg (eds), *Displacement, Diaspora, and Geographies of Identity*. Durham, NC: Duke University Press, pp. 157–79.

———. 2005. *Culture on Tour: Ethnographies of Travel*. Chicago: University of Chicago Press.

Burke, Patrick, Margareth Lanzinger, Daniela Salvucci, Peter Schöttler, Elisabeth Tauber, Lucy Ulrich, Lucie Varga, Pierpaolo Viazzo and Dorothy Louise Zinn. 2023. *Malinowski and the Alps – Anthropological and Historical Perspectives*. Bozen-Bolzano: bu,press.

Büscher, Bram, and Veronica Davidov. 2013. *The Ecotourism-Extraction Nexus: Political Economies and Rural Realities of (Un)Comfortable Bedfellows*. London: Routledge.

Caton, Kelle. 2012. 'Taking the Moral Turn in Tourism Studies'. *Annals of Tourism Research* 39(4): 1906–928.

Çağlar, Ayse. 2018. 'Chronotopes of Migration Scholarship: Challenges of Contemporaneity and Historical Conjuncture', in P. Barber and W. Lem (eds), *Migration, Temporality, and Capitalism*. Cham: Palgrave Macmillan.

Çağlar, Ayse, and Nina Glick Schiller. 2018. *Migrants and City-Making: Dispossession, Displacement, and Urban Regeneration*. Durham, NC: Duke University Press.

Candea, Matei. 2007. 'Arbitrary Locations: In Defence of the Bounded Field-Site', *Journal of the Royal Anthropological Institute* 13(1): 167–84.

Chambers, Erve. 1997. *Tourism and Culture: An Applied Perspective*. Albany: State University Press of New York.

Chaperon, Samantha, and Bill Bramwell. 2013. 'Dependency and Agency in Peripheral Tourism Development', *Annals of Tourism Research* 40: 132–54.

Chen, Jingfu, and Ningning Chen. 2017. 'Beyond the Everyday? Rethinking Place Meanings in Tourism', *Tourism Geographies* 19(1): 9–26.

Chio, Jenny T. 2014. *A Landscape of Travel: The Work of Tourism in Rural Ethnic China*. Seattle: University of Washington Press.

Clivaz, Christophe, Olivier Crevoisier, Leila Kebir, Stéphane Nahrath and Mathis Stock. 2014. 'Resort Development and Touristic Capital of Place', in Working Paper 5. Neuchâtel, Switzerland: University of Neuchâtel.

Cohen, Erik. 1979. 'A Phenomenology of Tourist Experiences', *Sociology* 13(2): 179–201.

———. 1986. 'Tourism and Time', *World Leisure & Recreation* 28(5): 13–16.

———. 1988. 'Authenticity and Commoditization in Tourism', *Annals of Tourism Research* 15(3): 371–86. https://doi.org/10.1016/0160-7383(88)90028-X.

Cole, John W., and Eric R Wolf. 1999. *The Hidden Frontier: Ecology and Ethnicity in an Alpine Valley*. Oakland: University of California Press.

Cretton, Viviane, Andrea Boscoboinik and Andrea Friedli. 2020. 'À l'aise, ici et ailleurs. Mobilités multirésidentielles en zone de montagne: le cas de Verbier en Suisse', *Anthropologie et Sociétés* 44(2): 107–26.

Crossley, Émilie. 2020. 'Temporality and Biography in Tourism: A Qualitative Longitudinal Approach', *Journal of Qualitative Research in Tourism* 1(1): 93–111.

Dahinden, Janine. 2016. 'A Plea for the "De-Migranticization" of Research on Migration and Integration', *Ethnic and Racial Studies* 39(13): 2207–25.

Dann, Graham. 1998. '"There's No Business Like Old Business": Tourism, the Nostalgia Industry of the Future', in W. Theobald (ed.), *Global Tourism*. Oxford: Butterworth Heinemann, pp. 29–43.

de Bernardi, Cecilia. 2019. 'A Critical Realist Appraisal of Authenticity in Tourism: The Case of the Sámi', *Journal of Critical Realism* 18(4): 437–52. doi: 10.1080/14767430.2019.1654702.

De Genova, Nicholas. 2016. 'The "Native's Point of View" in the Anthropology of Migration', *Anthropological Theory* 16(2–3): 227–40.

Denning, Andrew. 2014. *Skiing into Modernity: A Cultural and Environmental History*, Vol. 3. Oakland: University of California Press.

de Santana Pinho, Patricia. 2008. 'African-American Roots Tourism in Brazil', Latin American Perspectives 35(3): 70–86.

Duffy, Rosaleen. 2006. 'The Politics of Ecotourism and the Developing World', *Journal of Ecotourism* 5(1–2): 1–6.

Edensor, Tim, and Julian Holloway. 2008. 'Rhythmanalysing the Coach Tour: The Ring of Kerry, Ireland', *Transactions of the Institute of British Geographers* 33(4): 483–501.

Enzensberger, Hans Magnus. 1996. 'A Theory of Tourism', *New German Critique* 68: 117–35.

Eriksen, Thomas Hylland. 2016. *Overheating an Anthropology of Accelerated Change*. London: Pluto.

Erisman, H. Michael. 1983. 'Tourism and Cultural Dependency in the West Indies', *Annals of Tourism Research* 10(3): 337–61. https://doi.org/10.1016/0160-7383(83)90061-0.

Fabian, Johannes. 1983. *Time and the Other: How Anthropology Makes Its Object*. New York: Columbia University Press.

Ferguson, James. 2013. 'Declarations of Dependence: Labour, Personhood, and Welfare in Southern Africa', *The Journal of the Royal Anthropological Institute* 19(2): 223–42.

———. 2015. *Give a Man a Fish: Reflections on the New Politics of Distribution*. Durham, NC: Duke University Press.

Franklin, Adrian. 2008. 'The Tourism Ordering: Taking Tourism More Seriously as a Globalising Ordering', *Civilisations. Revue Internationale d'Anthropologie et de Sciences Humaines* 57: 25–39.
Frohlick, Susan. 2012. *Sexuality, Women, and Tourism: Cross-Border Desires through Contemporary Travel.* New York: Routledge.
Glick Schiller, Nina. 2021. 'Migration, Displacement, and Dispossession', in *Oxford Research Encyclopedia of Anthropology.*
Graburn, Nelson. 1977. 'Tourism: The Sacred Journey', in Valene L. Smith (ed.), *Hosts and Guests: The Anthropology of Tourism.* Philadelphia: University of Pennsylvania Press, pp. 17–32.
———. 1995. 'Tourism, Modernity and Nostalgia', in Akbar S. Ahmed and Chris N. Shore (eds), *The Future of Anthropology: Its Relevance to the Contemporary World.* London: Athlone, pp. 158–78.
———. 2002. 'The Ethnographic Tourist', in Graham Dann (ed.), *The Tourist as a Metaphor of the Social World.* Oxford: CABI, pp. 19–39.
Graf, Flurina. 2021. *Migration in den Alpen.* Bielefeld: Transcript-Verlag.
Greenwood, Davydd J. 1989. 'Culture by the Pound: An Anthropological Perspective on Tourism as Cultural Commoditization', in Valene L. Smith (ed.), *Hosts and Guests: The Anthropology of Tourism.* Philadelphia: University of Pennsylvania Press, pp. 171–85.
Guia, Aitana. 2016. 'The Concept of Nativism and Anti-immigrant Sentiments in Europe', *EUI Working Paper MWP 2016/20.*
Gupta, Akhil, and James Ferguson. 1992. 'Beyond "Culture": Space, Identity, and the Politics of Difference', *Cultural Anthropology* 7(1): 6–23.
Gyimóthy, Szilvia. 2018. 'The Indianization of Switzerland: Destination Transformations in the Wake of Bollywood Films', in *The Routledge Handbook of Popular Culture and Tourism,* pp. 376–87.
Hage, Ghassan. 2009. 'Waiting Out the Crisis: On Stuckedness and Governmentality', in Ghassan Hage (ed.), *Waiting.* Carlton, VIC: Melbourne University Press, pp. 97–106.
Haisch, Tina, Florian Jakob and Heike Mayer. 2014. 'Akteurspezifische Resilienz in ressourcenabhängigen Gemeinden: Wahrnehmung von Schocks und Anpassungsstrategien verschiedener Akteure in Grindelwald', in *CRED Bericht nr. 2.* Bern: Center for Regional Economic Development.
Haller, Tobias, Thomas Breu, Tine De Moor, Christian Rohr and Heinzpeter Znoj. 2019. *The Commons in a Glocal World: Global Connections and Local Responses.* Abingdon: Routledge.
Harvey, David. 2007. 'Neoliberalism as Creative Destruction', *Annals of the American Academy of Political and Social Science* 610(1): 21–44.
Hastrup, Kirsten. 1990. 'The Ethnographic Present: A Reinvention', *Cultural Anthropology* 5(1): 45–61.
Herzfeld, Michael. 2015. 'The Village in the World and the World in the Village: Reflections on Ethnographic Epistemology', *Critique of Anthropology* 35(3): 338–43.
Hirsch, Eric, and Charles Stewart. 2005. 'Introduction: Ethnographies of Historicity', *History and Anthropology* 16(3): 261–74.
Ioannides, Dimitri, and Kristina Zampoukos. 2018. 'Tourism's Labour Geographies: Bringing Tourism into Work and Work into Tourism', *Tourism Geographies* 20(1): 1–10. doi:10.1080/14616688.2017.1409261.

Jaeggi, Urs Josef Vitor. 1965. *Berggemeinden im Wandel: eine empirisch-soziologische Untersuchung in vier Gemeinden des Berner Oberlandes*. Bern: Haupt.

Jansen, Stef. 2015. *Yearnings in the Meantime: 'Normal Lives' and the State in a Sarajevo Apartment Complex*, Vol. 15. New York: Berghahn Books.

Janta, Hania, Lorraine Brown, Peter Lugosi and Adele Ladkin. 2011. 'Migrant Relationships and Tourism Employment', *Annals of Tourism Research* 38(4): 1322–43.

Janta, Hania, and Adele Ladkin. 2024. 'The Labour Consequences of Covid-19: Migrant Workers in Tourism and Hospitality', *Tourism Review* 79(1): 266–70.

Kaaristo, Maarja. 2018. 'Engaging with the Hosts and Guests: Some Methodological Reflections on the Anthropology of Tourism', in Sabina Owsianowska and Magdalena Banaszkiewicz (eds), *Anthropology of Tourism in Central and Eastern Europe: Bridging Worlds*. Lanham, MD: Rowman & Littlefield, pp. 71–88.

Kahn, Miriam. 2011. *Tahiti Beyond the Postcard: Power, Place, and Everyday Life (Culture, Place, and Nature)*. Seattle: University of Washington Press.

Kalbermatter, Jacqueline. 2020. *Bleiberecht in der Gastro-Küche: migrations-politische Regulierungen und Arbeitsverhältnisse von Geflüchteten mit unsicherem Aufenthaltsstatus*. Zurich: Seismo.

Kensbock, Sandra, Gayle Jennings, Janis Bailey and Anoop Patiar. 2016. 'Performing: Hotel Room Attendants' Employment Experiences', *Annals of Tourism Research* 56: 112–27. https://doi.org/10.1016/j.annals.2015.11.010.

Kilani, Mondher. 1984. 'Les images de la montagne au passé et au présent. L'exemple des Alpes valaisannes', *Schweizerisches Archiv für Volkskunde Basel* 80 (1–2): 27–55.

Kleist, Nauja, and Stef Jansen. 2016. 'Introduction: Hope over Time – Crisis, Immobility and Future-Making', *History and Anthropology* 27(4): 373–92.

Knight, Daniel M. 2015. *History, Time, and Economic Crisis in Central Greece*. New York: Palgrave Macmillan.

———. 2016. 'Temporal Vertigo and Time Vortices on Greece's Central Plain', *The Cambridge Journal of Anthropology* 34(1): 32–44.

———. 2017. 'Energy Talk, Temporality, and Belonging in Austerity Greece', *Anthropological Quarterly* 90(1): 167–91.

———. 2021. *Vertiginous Life: An Anthropology of Time and the Unforeseen*, Vol. 2. New York: Berghahn Books.

Knight, Daniel M., and Charles Stewart. 2016. 'Ethnographies of Austerity: Temporality, Crisis and Affect in Southern Europe', *History and Anthropology* 27(1): 1–18.

Krauß, Werner. 2018. 'Alpine Landscapes in the Anthropocene: Alternative Common Futures', *Landscape Research* 43(8): 1021–31. doi: 10.1080/01426397.2018.1503242.

Krippendorf, Jost. 1975. *Die Landschaftsfresser: Tourismus und Erholungslandschaft, Verderben oder Segen?* Bern, Stuttgart: Hallwag.

Ladkin, Adele. 2014. 'Labor Mobility and Labor Market Structures in Tourism', in Alan A. Lew, C. Michael Hall and Allan M. Williams (eds), *The Wiley Blackwell Companion to Tourism*. Wiley Online Library, pp. 132–42.

Lamont, Michèle, and Virág Molnár. 2002. 'The Study of Boundaries in the Social Sciences', *Annual Review of Sociology* 28(1): 167–95.

Leimgruber, Walter. 2005. 'Alpine Kultur: Welche Kultur für welchen Raum?', in Beate Binder, Silke Göttsch, Wolfgang Kaschuba, Komad Vanja (eds), *Ort. Arbeit. Körper. Ethnografie Europäischer ModernenI*. Munster: Waxmann, pp. 147–55.

———. 2019. ‘Constructing a Home: Heimat as Expression of Privilege, Belonging, Exclusion, and Identity’, *Journal for European Ethnology and Cultural Analysis* 1: 142–67.

Leite, Naomi. 2014. ‘Locating Imaginaries in the Anthropology of Tourism’, in Noel B. Salazaar and Nelson H.H. Graburn (eds), *Tourism Imaginaries: Anthropological Approaches*. New York: Berghahn Books, pp. 260–78.

———. 2017. *Unorthodox Kin: Portuguese Marranos and the Global Search for Belonging*. Oakland: University of California Press.

Leite, Naomi, Quetzil Castañeda and Kathleen Adams. 2019. ‘The Ethnography of Tourism: Encounter, Experience, Emergent Culture’, in Naomi Leite, Quetzil Castañeda and Kathleen Adams (eds), *The Ethnography of Tourism: Edward Bruner and Beyond*. Lanham, MD: Rowman & Littlefield.

Leite, Naomi, and Nelson Graburn. 2009. ‘Anthropological Interventions in Tourism Studies’, in Tazim Jamal and Mike Robinson (eds), *The SAGE Handbook of Tourism Studies*. London: Sage.

Leitenberg, Danaé. 2023a. ‘“In and Out of Time”: Towards an Anthropology of the Mundane Experiences of Modern and Capitalist Time’, *History and Anthropology* 35(4): 806–26. doi: 10.1080/02757206.2023.2204889.

———. 2023b. ‘Power Relations and Phenomenological Anthropology’, *Zeitschrift für Ethnologie / Journal of Social and Cultural Anthropology* 148(1): 119–22.

Lems, Annika. 2023. ‘Anti-mobile Placemaking in a Mobile World: Rethinking the Entanglements of Place, Im/mobility and Belonging’, *Mobilities* 18(4): 620–34. doi: 10.1080/17450101.2023.2220942.

Lerch, Ernst. 1909. ‘Die bernische Auswanderung nach Amerika im 18. Jahrhundert’, *Blätter für Bernische Geschichte, Kunst und Altertumskunde* 5(4).

Loloum, Tristan. 2015. ‘Derrière la plage, les plantations: ethnographie d’une ‘situation touristique’ dans le Nordeste brésilien: le cas de Tibau do Sul, RN’. Paris: EHESS.

López-González, José L., and David A. Fennell. 2020. ‘Social Acceleration as a New Frontier for the Ethics of Tourism’, *Recerca* 26(1): 1–7.

MacCannell, Dean. 2013. *The Tourist: A New Theory of the Leisure Class*. Berkeley: University of California Press.

Malkki, Liisa H. 1997. ‘News and Culture: Transitory Phenomena and the Fieldwork Tradition’, in Akhil Gupta and James Ferguson (eds), *Anthropological Locations: Boundaries and Grounds of a Field Science*. Berkeley: University of California Press, pp. 86–101.

Marcher, Anja, Ingrid Kofler, Elisa Innerhofer and Harald Pechlaner. 2022. ‘Perceptions of and Interactions between Locals, Migrants, and Tourists in South Tyrol’, *Tourism Geographies* 24(1): 56–72. doi: 10.1080/14616688.2020.1795709.

Martin, Keir. 2021. ‘Dependence’. *Cambridge Encyclopedia of Anthropology*.

Martin, Keir, Ståle Wig and Sylvia Yanagisako. 2021. ‘Battlegrounds of Dependence: Reconfiguring Labor, Kinship and Relational Obligation’, *Focaal* 90: 1–10.

Martin, Keir, and Sylvia Yanagisako. 2020. ‘States of Dependence: Introduction’, *Social Anthropology* 28(3): 646–56.

McDowell, Linda, Adina Batnitzky and Sarah Dyer. 2007. ‘Division, Segmentation, and Interpellation: The Embodied Labors of Migrant Workers in a Greater London Hotel’, *Economic Geography* 83(1): 1–25.

———. 2009. ‘Precarious Work and Economic Migration: Emerging Immigrant

Divisions of Labour in Greater London's Service Sector', *International Journal of Urban and Regional Research* 33(1): 3–25.

Membretti, Andrea, and Pier Paolo Viazzo. 2017. 'Negotiating the Mountains: Foreign Immigration and Cultural Change in the Italian Alps', *Martor* 22: 93–107.

Messerli, Paul. 1987. 'The Development of Tourism in the Swiss Alps: Economic, Social, and Environmental Effects Experience and Recommendations from the Swiss MAB Programme', *Mountain Research and Development* 7(1): 13–23.

Milano, Claudio. 2018. 'Overtourism, Social Unrest and Tourismphobia: A Controversial Debate', *PASOS: Revista de Turismo y Patrimonial Cultural* 16(3): 551–64.

Milano, Claudio, Marina Novelli and Joseph M. Cheer. 2019. 'Overtourism and Tourismphobia: A Journey Through Four Decades of Tourism Development, Planning and Local Concerns', *Tourism Planning & Development* 16(4): 353–57. doi: 10.1080/21568316.2019.1599604.

Milano, Claudio, Marina Novelli and Antonio Paolo Russo. 2024. 'Anti-tourism Activism and the Inconvenient Truths about Mass Tourism, Touristification and Overtourism', *Tourism Geographies* 26(8): 1313–37. doi: 10.1080/14616688.2024.2391388.

Moore, Amelia. 2019. *Destination Anthropocene: Science and Tourism in the Bahamas*, Vol. 7. Oakland: University of California Press.

Morgan, Lynn M. 1987. 'Dependency Theory in the Political Economy of Health: An Anthropological Critique', *Medical Anthropology Quarterly* 1(2): 131–54.

Moroşanu, Roxana, and Felix Ringel. 2016. 'Time-tricking: A General Introduction', *The Cambridge Journal of Anthropology* 34(1): 17–21.

Moser, Patrick. 1997. *'So wird die Jungfrau zur Demoiselle gemacht': Projektierung und Bau der Jungfraubahn*. Zurich: Chronos.

Mostafanezhad, Mary, Matilde Córdoba Azcárate and Roger Norum. 2021. *Tourism Geopolitics: Assemblages of Infrastructure, Affect, and Imagination*. Tucson: University of Arizona Press.

Mostafanezhad, Mary, and Kevin Hannam. 2016. *Moral Encounters in Tourism*. London: Routledge.

Mughal, Muhammad Aurang Zeb. 2015. 'Being and Becoming Native: A Methodological Enquiry into Doing Anthropology at Home', *Anthropological Notebooks* 21(1): 121–32.

Narotzky, Susana, and Niko Besnier. 2014. 'Crisis, Value, and Hope: Rethinking the Economy: An Introduction to Supplement 9', *Current Anthropology* 55(S9): S4–S16. doi: 10.1086/676327.

Nader, Laura. 1972 [1969]. 'Up the Anthropologist: Perspectives Gained from Studying Up', in Dell Hymes (ed.), *Reinventing Anthropology*. New York: Random House, pp. 284–311.

Nash, Dennison. 1989. 'Tourism as a Form of Imperialism', in Valene L. Smith (ed.), *Hosts and Guests: The Anthropology of Tourism*. Philadelphia: University of Pennsylvania Press, pp. 37–52.

Nathani, Carsten, Julia Brandes, Beat Burri, and Tonio Schwehr. 2022. *Die Bedeutung des Gastgewerbes für Wirtschaft und Arbeitsmarkt in der Schweiz: Schlussbericht an GastroSuisse*. Rüschlikon: Rütter Soceco AG.

Netting, Robert M. 1981. *Balancing on an Alp: Ecological Change and Continuity in a Swiss Mountain Community*. Cambridge: Cambridge University Press.

Niederer, Arnold, Werner Bätzing and Klaus Anderegg. 1993. *Alpine Alltagskultur zwischen Beharrung und Wandel: ausgewählte Arbeiten aus den Jahren 1956 bis 1991.* Bern: Paul Haupt.

Nöbauer, Herta. 2022. 'Our Existence Is Literally Melting Away', in Susanna M. Hoffman, Eriksen Thomas Hylland and Mendes Paulo (eds), *Cooling Down: Local Responses to Global Climate Change.* New York: Berghahn Books, pp. 223–44.

Norum, Roger, and Mary Mostafanezhad. 2016. 'A Chronopolitics of Tourism', *Geoforum* 77: 157–60.

Peeters, Paul, et al. 2018. 'Research for TRAN Committee – Overtourism: Impact and Possible Policy Responses'. A European Union Study.

Petite, Mathieu. 2013. 'Les montagnards face aux écolos. Une énième déclinaison au travers de l'initiative Weber', *Journal of Alpine Research* (Hors-Série).

Picard, David, and Mike Robinson. 2012. *Emotion in Motion: Tourism, Affect and Transformation.* Burlington, VT: Ashgate.

Purtschert, Patricia, Barbara Lüthi and Francesca Falk. 2012. *Postkoloniale Schweiz. Formen und Folgen eines Kolonialismus ohne Kolonien.* Bielefeld: transcript.

Ramsay, Georgina. 2019. 'Time and the Other in Crisis: How Anthropology Makes Its Displaced Object', *Anthropological Theory* 4: 385–413.

Ringel, Felix. 2016. 'Beyond Temporality: Notes on the Anthropology of Time from a Shrinking Fieldsite', *Anthropological Theory* 16(4): 390–412.

———. 2018. *Back to the Postindustrial Future: An Ethnography of Germany's Fastest-Shrinking City.* New York: Berghahn Books.

Rosa, Hartmut, Klaus Dörre and Stephan Lessenich. 2017. 'Appropriation, Activation and Acceleration: The Escalatory Logics of Capitalist Modernity and the Crises of Dynamic Stabilization', *Theory, Culture & Society* 34(1): 53–73.

Salazar, Noel B., and Nelson H.H. Graburn. 2014. *Tourism Imaginaries: Anthropological Approaches.* New York: Berghahn Books.

Santos, Priscilla, Malet Calvo, Daniel, and Nofre, Jordi (2025). 'Navigating between Resistance and Unintentional Collaboration: The Role of Left-wing Grassroots Associations in the Tourist City'. *Urban Studies* 62(4): 737–753.

Saxer, Martin and Ruben Andersson. 2019. 'The Return of Remoteness: Insecurity, Isolation and Connectivity in the New World Disorder', *Social Anthropology* 27(2): 140–155.

Schär, Bernhard. 2012. 'Bauern und Hirten Reconsidered: Umrisse der "erfundenen" Schweiz im imperialen Raum', in Patricia Purtschert, Barbara Lüthi and Francesca Falk (eds), *Postkoloniale Schweiz. Formen und Folgen eines Kolonialismus ohne Kolonien.* Bielefeld: transcript, pp. 315–31.

Schmid, Samuel. 2023. 'Democratising Switzerland: The Significance of the New Naturalisation Initiative', *EUI Global Citizenship Observatory,* Florence: Globalcit, https://globalcit.eu/democratizing-switzerland/.

Schelbert, Leo. 1976. *Einführung in die schweizerische Auswanderungsgeschichte der Neuzeit.* Zurich: Stäubli.

Schneider, Almut, and Elisabeth Tauber (eds). 2024. *New Horizons for the Alps: Ethnographies, Reshaping Challenges, and Emerging More-Than-Alpine Relations.* Bozen-Bolzano: bu,press.

Schuurman, Frans J. 1993. 'Introduction: Development Theory in the 1990s', in Frans J. Schuurman (ed.), *Beyond the Impasse: New Directions in Development Theory.* London: Zed Books, pp. 1–48.

Seiger, Fiona-Katharina, Christiane Timmerman, Noel B. Salazar and Johan Wets (eds). 2020. *Migration at Work: Aspirations, Imaginaries & Structures of Mobility*, Vol. 5. Leuven: Leuven University Press.

Sherlock, Kirsty. 2001. 'Revisiting the Concept of Hosts and Guests', *Tourist Studies* 1(3): 271–95. doi: 10.1177/146879760100100304.

Sidali, Katia Laura, and María de Obeso. 2017. 'Successful Integration of Slow and Sustainable Tourism: A Case Study of Food Tourism in the Alpine Region of Algovia, Germany', in Michael Clancy (ed.), *Slow Tourism, Food and Cities*. Abingdon: Routledge, pp. 169–80.

Simoni, Valerio. 2018. 'Approaching Difference, Inequality, and Intimacy in Tourism: A View from Cuba', *Journal of Anthropological Research* 74(4): 503–25.

———. 2019. 'Living in and Reaching beyond the Touristic Borderzone: A View from Cuba', in Naomi Leite, Quetzil Castañeda and Kathleen Adams (eds), *The Ethnography of Tourism: Edward Bruner and Beyond*. Lanham, MD: Rowman & Littlefield, pp. 107–24.

Smith, Valene (ed.). 1989. *Hosts and Guests: The Anthropology of Tourism*. Philadelphia: University of Pennsylvania Press.

Stacul, Jaro. 2003a. *The Bounded Field: Localism and Local Identity in an Italian Alpine Valley*, Vol. 18. New York: Berghahn Books.

———. 2003b. 'The Privatisation of Locality: Integralist Political Engagements in Northern Italy at the Turn of the Millennium', *Cambridge Anthropology* 23(2): 36–53.

———. 2006. 'Neo-nationalism or Neo-localism? Integralist Political Engagements in Italy at the Turn of the Millennium', in Marcus Banks and Andre Gingrich (eds), *Neo-nationalism in Europe and Beyond: Perspectives from Social Anthropology*. New York: Berghahn Books: 162–76.

Starobinski, Jean. 1966. 'The Idea of Nostalgia', *Diogenes* 14(54): 81–103.

Stasch, Rupert. 2017. 'Tourism', in *The Cambridge Encyclopedia of Anthropology*, edited by Feliz Stein, Sian Lazar, Matei Candea, Hildegard Diemberger, Joel Robbins, Andrew Sanchez and Rupert Stasch.

Stephen, Leslie. 1895. *The Playground of Europe*. London: Longmans, Green and Co.

Stone, Alison. 2014. 'Alienation from Nature and Early German Romanticism', *Ethical Theory and Moral Practice* 17(1): 41–54.

Stotten, Rike, Markus Schermer and Geoff A. Wilson. 2021. 'Lock-Ins and Community Resilience: Two Contrasting Development Pathways in the Austrian Alps', *Journal of Rural Studies* 84: 124–33. doi: https://doi.org/10.1016/j.jrurstud.2021.04.001.

Stronza, Amanda. 2005. 'Hosts and Hosts: The Anthropological of Community-Based Ecotourism in the Peruvian Amazon', *NAPA Bulletin* 23(1): 170–90.

Tiefenbach, Marianne. 2013. *Alpkorporationen – traditionelle Institutionen nachhaltiger Landschaftsentwicklung: das Beispiel der Bergschaften Grindelwalds im Kontext aktueller gesellschaftlicher Herausforderungen*. Grindelwald: Sutter Druck.

Tissot, Laurent. 2004. 'Le tourisme en Suisse ou l'avènement d'un modèle d'excellence (19e-20e siècles)', *Le Globe. Revue genevoise de géographie* 144(1): 103–21.

Tsing, Anna Lowenhaupt. 2015. *The Mushroom at the End of the World: On the Possibility of Life in Capitalist Ruins*. Princeton, NJ: Princeton University Press.

Tucker, Hazel. 2003. *Living with Tourism: Negotiating Identities in a Turkish Village*: London: Routledge.

Turner, Louis. 1976. 'The International Division of Leisure Tourism and the Third World', *Annals of Tourism Research* 4(1): 12–24.

Urry, John, and Jonas Larsen. 2011. *The Tourist Gaze 3.0*. London: Sage.

van den Berghe, Pierre L. 1992. 'Tourism and the Ethnic Division of Labor', *Annals of Tourism Research* 19(2): 234–49.

Viazzo, Pier Paolo. 1989. *Upland Communities: Environment, Population and Social Structure in the Alps since the Sixteenth Century*, Vol. 8. Cambridge: Cambridge University Press.

Vietti, Francesco. 2024. *Unexpected Encounters: Migrants and Tourists in the Mediterranean*. Vol. 4. New York: Berghahn Books.

Vorhölter, Julia. 2021. 'Anthropology Anonymous? Pseudonyms and Confidentiality as Challenges for Ethnography in the Twenty-first Century', *Ethnoscripts* 23(1): 15–33.

Wang, Ning. 2000. *Tourism and Modernity: Sociological Analysis*. 1st edn. Vol. 5, *Tourism Social Science Series*. Amsterdam: Pergamon.

Wanner, Alice, Ulrike Pröbstl-Haider and Magdalena Feilhammer. 2021. 'The Future of Alpine Pastures – Agricultural or Tourism Development? Experiences from the German Alps', *Journal of Outdoor Recreation and Tourism* 35: 100405.

Watts, Galen, and Dick Houtman. 2024. 'The Specter of Authenticity: Social Science after the Deconstruction of Romanticism', *History of the Human Sciences* 37(5): 45–67. doi: 10.1177/09526951241245014.

Wiesmann, Urs, and Karina Liechti. 2004. 'The Contributions of World Natural Heritage Sites to Sustainable Regional Development – Two Case Studies from the North and the South', *Revue de géographie alpine* 92(3): 84–94.

Williams, Stephen, and Alan A. Lew. 2014. *Tourism Geography: Critical Understandings of Place, Space and Experience*. London: Routledge.

Wimmer, Andreas. 2011. 'A Swiss Anomaly? A Relational Account of National Boundary-Making', *Nations and Nationalism* 17(4): 718–37. doi: https://doi.org/10.1111/j.1469-8129.2011.00517.x.

Yuval-Davis, Nira. 2006. 'Belonging and the Politics of Belonging', *Patterns of Prejudice* 40(3): 197–214.

Zimmer, Oliver. 1998. 'In Search of Natural Identity: Alpine Landscape and the Reconstruction of the Swiss Nation', *Comparative Studies in Society and History* 40(4): 637–65.

CHAPTER

1

FROM FARMING TO THE BUSINESS OF FOREIGNERS

'The world comes to visit us here', says Kristi, a local sports instructor and employee of a touristic company in her thirties from Zurich. It is why she chose to move here decades ago. Two weeks into my fieldwork, she tells me this as we are standing in front of her massage studio, near the queues of tourists on the busy *Hauptstrasse*, who have travelled from China, India or Saudi Arabia to visit one of the most visited sightseeing spots in the valley. Looking at them, Kristi insists that this is the stuff that makes the valley a 'very multicultural place'. Tourism brings a particular, urban flair to the village – an atmosphere one could not get in the more isolated surrounding areas. The village, Kristi tells me, is connected to the outside, to the urbanized lowlands stretching from the nearby town of Interlaken to Zurich, and abroad. It is globalized in other words, and 'people here are open-minded you know, thanks to tourism'. She pauses. 'I mean not *all* the villagers are', but she likes all the cultures here.

A few months later, I make my way through the crowds of tourists from all over the world to conduct an interview with Moritz. He lives on the outskirts of the village, among houses scattered across pastures (the *Streusiedlung*). Moritz drove down to meet me in the centre – where each centimetre of land is covered by roads, buildings, houses – in one of the village's hotel restaurants, bordering the bustling *Hauptstrasse*. I assume it is his 4×4 parked in front of the hotel, as I see the sticker on his windshield that allows him, like other legal first residents willing to pay for it, to use the roads of the valley that are barred to non-locals. I have not met him yet but I can already picture what Moritz is going to look like, based on how other villagers have described him: a typical 'native' (*Einheimische*), a stubborn mountain dweller (*Bergler*), a freedom lover, a rebel. When I told locals that I was studying the valley, some insisted that I interview him, to get a sense of how the *genuine* villagers 'ticked'. Upon entering the restaurant, Moritz stands up from one of the few *Stammtisch* (the regulars' table) of the village with other older men, all wearing working clothes, hats with the names of local clubs, and dirty hiking shoes – 'the *Einheimische* outfit', I think to myself. As he joins me, I can feel the inquisitive gazes of his peers, who are clearly trying to understand why he would possibly meet with me, a woman half his age with a French accent and dressed in urban clothes.

After sitting down, Moritz warns me he will talk 'unapologetically' as he has a lot to say about the village and its development, often not in a positive sense. The *Touristikers* – the tourism lobbyists – would probably dislike his words, he says, but he does not care. Moritz says he belongs to the few, disappearing *Einheimische* or even *Ureinheimische* (genuine natives), as his authentic local family name also indicates and is proud to call the village his *Heimat* – his homeland, the centre of his world. The valley is the place where his father, grandfather and great-grandfather came from, where he inherited a house and then bought a new one thanks to local networks. It is the place where he hunts, gathers mushrooms, and works as a carpenter in summer and runs a small business selling local food specialties in winter. His hands are those of a hard-working man, of someone who spends a lot of time outside, preferably up the valley's sides, fixing, repairing, walking, hunting – activities I heard in the background when we phoned to arrange meetings.

During our interview, Moritz tells me that he sees his village as an isolated valley, one of its kind, inhabited by genuine, special and unique mountain people, shaped by a harsh environment and a particular history of farmers' resistance against elites and foreign bailiffs from the 'outside'. He refers to how his farming ancestors fought against the domination of the Covent of Interlaken under, among others, Habsburgian power and their forced conversion to Protestantism with the Reformation in 1528. Moritz says that this spirit of resistance is still alive here, in the village, but also Switzerland, exemplified in the refusal to join the EU and in a commitment to conservatism and defending Swiss and local values against those of the outside, from abroad, or the 'red', leftist lowlands. 'The freedom to decide here is still greater than elsewhere', he adds. But it is fading away. Immigration, urbanization, technological advancements and increasing bureaucratization and centralization, along with mass tourism, all weigh heavily on Moritz's valley, the country and his people, who, instead of being given 'priority', are increasingly falling victim to external influence.

Kristi's cosmopolitan ideals were shared by many locals, usually originally from the 'outside' of the village. Listening to them, living off tourism logically led one to cherish and enjoy the presence of people from elsewhere in the valley. If travelling the world as a tourist evokes the cosmopolitan activity par excellence (Salazar and Graburn 2014), Kristi and her peers seemed to derive a similar, enlightened, cosmopolitan sensibility from their tourism dependency. Their opinions reflected many of those circulating in my own progressive environment and I, like them, had often mulled over how people like Moritz could share such a radically different vision, in spite of the fact that he, too, lived off tourism and globalization. How could he be so against something he so clearly seemed to profit from?

If all villagers acknowledged their reliance on globalization and international tourism, a great number of locals like Moritz stressed their right to criticize and question these processes, precisely because they were involved in them. At a time of resurgence of far-right and antimodern movements globally (Mazzarella 2019; Pasieka 2019), many locals in the valley shared the global scepticism towards universalist narratives of development, knowledge or belonging, which they saw as threatening and/or divorced from their everyday realities. How, then, were these antimodern reactions and worldviews to be grasped in relation to an obviously globalized present? This first chapter sets out to respond to this question by exploring the relationship between modernity and antimodernity, cosmopolitanism and anticosmopolitanism, while tracing their deep-rooted origins in the valley. It sheds light on the empirical reality of tourism development, its long history in the German-speaking Swiss Alps, and the ways in which modernity and capitalism frame such development in linear and entirely beneficial ways for local populations but thereby obfuscate the messiness and unevenness of tourism development. As I will argue here, the problem with such a modern framing is that it overlooks more critical, anticosmopolitan and nativist sensibilities, which have also historically shaped this region until nowadays and the Romantic ideals at the very origins of tourism. As the first chapter of a village ethnography, the following pages make up a history chapter, albeit not an entirely classical one: while it traces the village's historical development over the last centuries, it also examines how the past and its relationship to the present serve modern and antimodern narratives of development, connection and belonging for my interlocutors.

TOURISM AND THE MODERNIST TELEOLOGY

When I told them that I was interested in the history of the valley, tourists, holiday-home owners, Swiss and/or *Einheimische* locals and migrant hospitality workers unanimously encouraged me to picture the past of the village as a poor and isolated farming community that had succumbed to modernity and progress thanks to tourism. This was not entirely unimaginable: in the hotels and in the restaurants, I was confronted by the omnipresence of romanticized pictures depicting the hard but enjoyable life of farmers in the Alpine pastures. These pictures, often recently made, displayed farmers working with rakes and sickles instead of the special and expensive hay mowers they use nowadays.

The more informed villagers drew on less idealized historical depictions of children begging on the village streets at the end of the nineteenth century, or of the massive outmigration of local farmers to North America, presented in various works on the Bernese Alps (e.g. Schelbert 1976; Rubi 1993: 53). Sometimes, this past was equated with the current state of an 'elsewhere' – in the mountains of Nepal, the Himalayas, or Africa. Amidst crowds and success, the people I met urged me to imagine how there was nothing here, how everyone was poor and disconnected from the rest of the world. Tourism, they contended, had shaped the emergence of an 'after'. Locals pointed to the declining number of farmers since 1900, highlighting the inevitability of modernization via touristification. In their libraries, books attested to an unstoppable march of history, in which tourism played a central role in catapulting the poor native farmers of the region – whose lives were featured by 'little or no change' – into modernity. Tourism turned the mountain dwellers' small world upside down, into one of intense connections and business with 'strangers' (Rubi 1986).[1]

Plenty of contemporary and historical empirical data indeed reveals tourism's undeniable, historical and contemporary effects on the local socio-economic structure of the Alps (Niederer, Bätzing and Anderegg 1993; Debarbieux, Del Biaggio and Petite 2008) and beyond (Sampaio, Simoni and Isnart 2014). Many tourism theorists even equate the emergence of modern tourism in the Alps with the arrival of European Romantics in the eighteenth century, after a long period of disregard, fear and disdain for mountainous areas (Enzensberger 1996; Mathieu and Boscani Leoni 2005). Opposed to the rationalization and industrialization processes permeating Europe at the time, these intellectuals and artists – usually members of elites – found in the Alps a place of purity, preserved from the moral and material deterioration they denounced in urbanized lowlands (Starobinski 1966). Jean-Jacques Rousseau and Felix Mendelssohn translated their experiences of the Alps into a Romantic vocabulary, describing their encounters with the sublime Nature as healing and de-alienating (Bergier et al. 2013). They celebrated not only the locales but also the people that inhabited them. In Switzerland in particular, historian Bernard Schär has shown how Swiss scientist and poet Albrecht von Haller viewed mountain dwellers as poor but honest, naive but noble Alpine 'peasants and shepherds' (*Bauern und Hirten*), like those I saw in the pictures displayed in the village's hotels.[2]

In time, as I sketched in the Introduction of this book, the newly emerging capitalist industry of 'tourism'[3] successfully captured such critical sensibilities and aesthetics to commodify an escape from the everyday to an increasing number of travellers, inspired by the Romantics' tales (Denning 2014). Mountaineers, tourists on Thomas Cook's famous organized tours, sick people seeking pure air, Tsarist and British aristocrats, writers and painters, skiers, holiday-home owners and group tourists had all visited the valley for its extraordinariness, its 'exceptional' environment, over the past centuries. Its glaciers offered memorable views, with their giant ice tongues traversing the valley's green pastures in the eighteenth and nineteenth centuries, before they melted. The peaks of the valley's surrounding mountains, such as the *Dreigestirn*, attracted many climbers. A large infrastructure progressively deployed in the Alps as roads, a railway system, hotels and restaurants mushroomed in previously remote valleys, allowing for the large-scale democratization of tourist experiences at the end of the nineteenth century (Tissot 2006; Barton 2008).

In a valley known for its exceptionally long and stable history of touristification, dating the construction of roads, railways and lifts criss-crossing the valley, or that of hotels, shops, restaurants, houses, museums, and their subsequent transformation over the years, was common practice among locals. They described the present as the result of past investments and inventions, where 'everything' was owed to tourism and the smooth, linear development it had enabled in the valley. But while all narratives pointed to tourism's overwhelming power in making the valley move forward, progressing and developing from a poverty-stricken and isolated past, I could not help but feel dissatisfied with this interpretation. Much of what locals explained and experienced, as well as further historical data on the valley and the Alps I consulted, did not seem to 'fit' – or even contradicted – the narrative of linear tourism development.

In his work on Zambia's economic decline from the mid-1970s into the 1990s, anthropologist James Ferguson provides a groundbreaking analysis of modernity, along with its ideology and effects on social actors. He refers to a 'linear metanarrative of emergence and progress' (Ferguson 1999: 16) shaping his interlocutors' expectations and their disappointment as these remain unfulfilled and inaccessible. For Ferguson, this proposes dichotomies between tradition and modernity, backward and progressive, isolated and connected, primitive and civilized, us and them, which are put in a temporal relation. The past, of poverty, rurality and isolation, is believed to ineluctably develop into the urbanized, cosmopolitan and generally 'better' future, which constitutes the goal – the *telos* – of a 'modernist teleology'.

But such smooth, linear narratives are in fact 'myths' in Ferguson's words: they force frames of interpretation on historical events and thus often distort, bend a reality that is full of frictions and non- or counterlinearities. However, the mythical aspects of modernity, Ferguson states, cannot just be discarded as inaccurate for they have social consequences; they shape the expectations and experiences for people who live under the temporal ideology of modernity.

My field site does not escape to this reality with the tale *from* timelessness, disconnection, poverty and isolation *to* timeliness, connection and wealth; *from* an Alpine farming society *to* a tourism society; *from* a native's place *to* a foreigner's place, retold by locals but also thematized by a great number of works on the Jungfrau Region.[4] But there is plenty that the modern myth omits in this valley. Against the isolation and boundedness with which local historians (and Romantics) described farming 'natives' inhabiting the valley and the Alps before tourism 'impacted' the valley and its community, various data have attested to a reality of intra- and intercontinental migration. Swiss Alpine dwellers were rented out to the European powers as mercenaries, which took them far from their homelands as early as the seventeenth century (Netting 1981; Lems 2016; Haller et al. 2019: 2–3). Alpine dwellers also long resorted to emigration, more or less permanently, to other regions of the Alpine arc, as well as to Europe and other parts of the world, in order to gain stability, power and resources (Schelbert 1976; Viazzo 1989; Zanini 2013). Distant ancestors of locals in the valley, too, were highly mobile, hailing from what is now the neighbouring canton of Wallis, by crossing a pass that then became a glacier (Wiener 2022).

Poverty did not, contrary to the modern myth, vanish with the rise of tourism either. Historian Rudolf Rubi, himself one of the important actors shaping this narrative in the Jungfrau Region, reported that begging children did not predate tourism but had coexisted with it at the end of the nineteenth century (Rubi 1993: 153). Some villagers I talked to, especially from poorer, farming backgrounds, also recalled the seasonal labour they performed in the Swiss lowlands in the 1950s, either on farms or preparing ice-skating rinks and ski slopes. The wealth brought by tourism, many locals reported, did not reach every household equally in the originally more dispersed 'scattered settlement' (*Streusiedlung*) that made the valley, echoing with Jenny Chio's interlocutors complaints about the uneven distribution of wealth accompanying tourism development (Chio 2014).[5] This led multiple dwellers to leave the valley long after the beginning of its touristification. Locals I met who lived on the margins of the village, for instance, remembered times of being 'significantly poorer' than their counterparts, revealing a divide between hoteliers and farmers that continued to shape the valley during their youth, forty or fifty years prior to our meeting. At the same time, tourism development, if crucial, was marked by an overwhelming number of crises – not only in the present, as we shall see in future chapters, but also historically. The two world wars, the financial crash and the succession of financial crises since the 1970s, as well as more recently the terror attacks and epidemic and pandemic outbreaks, in 2009 and 2020 respectively, profoundly impacted the tourism industry and the village. In some cases, these crises even provoked a 'return' to farming, as had been the case during the First World War.[6]

For nearly all of my interlocutors, the shift to tourism by farmers in the valley seemed self-explanatory. Tourism was the better alternative, providing new forms of subsistence in a context shaped by poverty and environmental hostility. Yet, this shift clashed not only with local memories of a poverty coexisting with tourism but also with historical accounts revealing the contested and frictional nature of touristification. Coachmen of the region, for instance, wished to counter the advent of transport infrastructure because they feared for their subsistence (Rubi 1987; Moser 1997: 33–50). Like them, many village dwellers were more or less forced to adapt to the advent of the new industry of tourism if they wished to continue making a living.

In one of his books, historian Rubi describes such instances of resistance to the advent of the expansion of the railway network in the Jungfrau Region, but retrospectively regards these attitudes as irrational because locals had then 'benefited' from tourism development and its infrastructure (Rubi 1987: 40–41). However, much more than misjudgement, these historical facts reveal the omnipresent forms of scepticism and angst involved in each step of the touristification process in the valley. This still held true during my research. If tourism had effectively come to represent the only viable life in the Alps, this did not prevent many of the locals I spoke with from acknowledging its ambiguous, uncertain and uneven powers.

GLOBALIZATION, COSMOPOLITANISM AND COSMOPOLITANIZATION

The modern story of acceleration and connection that has accompanied tourism development in the valley is undoubtedly one of globalization. Herta Nöbauer speaks of an Alpine 'vertical globalization', describing 'the increasing flows of people, ideas, infrastructure, communication technology, trade and finance oriented towards high environmental elevations, high mountain areas and the sky' (Nöbauer 2021: 22). This process extended far beyond the confines of the valley and Switzerland, and brought about important changes for farmers, who progressively took up activities catering to the needs of tourists from different backgrounds, such as Alpine guides for foreign mountaineers. From the 1950s until the present, they also took up positions operating ski lifts and gondolas in winter, in combination with, or as a replacement for, their farming activities.

In the valley I conducted fieldwork, as in much of the (Swiss) Alps, 'vertical globalization' also involved demographic change (Viazzo 1989). Tourists of all kinds, and for periods of differing lengths, stayed in the village's hotel rooms, rented apartments, and sometimes even built or bought holiday houses.[7] In parallel, the tourism industry relied on the arrival of a workforce from the 'outside', crossing cantonal and national borders, to build and support the valley's important infrastructure (Kuhn 2019). During my visits to the communal archive, I found records of the presence of workers from Italy and Germany employed in the village's hotels as waiters, cooks and chambermaids from the end of the nineteenth century. The touristic economy historically relied on 'outsiders' (*Auswärtige*) and 'foreigners' (*Ausländer*) as the local workforce could not provide enough services for the countless tourists crossing the borders of the municipality. Just before my arrival in the valley, a house belonging to a municipality from the Bernese lowlands, destined to host school pupils during ski camps, had also been repurposed into a refugee centre for unaccompanied minors, mostly from Afghanistan and Eritrea.[8] In 2022, one of the village's military barracks, previously transformed into a hostel, was also opened to host refugees from Ukraine.[9]

The 'multicultural' reality that Kirsti presented when I met her was indeed a long-standing feature of the valley. But she did not only express her cosmopolitan values of openness and welcome to travellers from different backgrounds visiting and living in the valley during our discussion. She also defined cosmopolitanism as a form of teleology: because they had been living off globalized tourism, she believed that locals should have been open-minded. In a similar vein, other locals emphasized the development of a cosmopolitan flair of the formerly isolated, poor farmers' valley when it became a place visited by strangers (*Fremdenort*). A long history of globalization was thought (or expected) to have generated a form of enlightened, progressive attitude in these narratives, shared by plenty of villagers and tourism officials, who regularly praised local hospitality. The belief in universality and a general feeling of openness towards fellow human beings from other cultures, or, in other words, cosmopolitanism,[10] was supposed to naturally follow tourism development and globalization.

As much as the myth of modernity naturalizes processes of development, it also often presupposes the progressive dispositions of those embedded in these transformations. To be sure, engaging in increasing transnational connections nowadays constitutes an unavoidable part of life and, in many ways, progress. In the valley, a touristic resort saturated with cosmopolitan imaginaries of hospitality, and reliant on global connections to grow, relationships to worlds located beyond it were seen as particularly vital for my interlocutors. But then, how did people like Moritz, who lived in this globalized, touristic reality, remain deeply influenced by anticosmopolitan values?

Sociologist Ulrich Beck makes an important distinction between processes and attitudes that are often conflated: the reality of cosmopolitanization and the worldviews pertaining to cosmopolitanism. In cosmopolitanization, 'clear borders, separating markets, states, civilizations, religions, cultures, life-worlds of common people' collapse and shape a 'global order' in which individuals are 'included', for good or bad (Beck 2011: 57). For Beck, cosmopolitanization does not emerge from a 'conscious affirmation of cosmopolitan values'; instead, it constitutes a 'secondary side effect' of interconnection and flows of goods and people (ibid.: 58). He thereby distinguishes the philosophical understandings of cosmopolitanism from the more empirically grounded, 'involuntary' aspects involved in cosmopolitanization. As a result, Beck argues, those who inhabit cosmopolitanized places are not always cosmopolites, quite the contrary.

This rings true for various contexts around the globe, which have been intensively cosmopolitanized but where anticosmopolitan movements have recently (re-)gained popularity. It also mirrors the reality of my field site, as Kirsti even noted: while *she* enjoyed the multicultural flair of the valley, she knew that not all villagers did. Rather than the product of a teleology of touristic development, Kirsti's cosmopolitan sensibility appeared to stem from her personal qualities of openness and her appreciation of other cultures. Like many other locals I met, she chose to live in the valley because she saw in it a globalized place where such sensibility would find ways to express itself. In other words, contrary to what she seemed to imply, her cosmopolitanism did not derive from her tourism dependency as an inhabitant of a touristic resort, but found its origins elsewhere – in her education, her networks and her past experiences. By resisting the correlation between cosmopolitanism and tourism dependency, a link so often naturalized as a logical reaction in the teleology of modernity, being cosmopolite and anticosmopolite emerge as two possible sensibilities within the same cosmopolitanized or globalized reality – a reality where both Kirsti and Moritz could express radically different views on the village, in spite of their common reliance on tourism.

ANTIMODERNISM AND ANTICOSMOPOLITANISM IN THE GLOBAL VILLAGE

Moritz, like many of his native peers, did not deny the cosmopolitanized character of his *Heimat*; he knew what he was living off. He could see in his family how tourism had progressively changed the lifestyles of various generations, as it made for a life of more security and new possibilities. The past was that of poverty, the present that of tourism. Yet, something had been retained from the past: the isolated topography of the place, the mountainous geography and conservative values derived from farming, all shaped the local natives' identities and positions in the present – 'thankfully', according to Moritz, who believed that this past-oriented attitude was a prerequisite for the preservation of 'freedom' in a globalized world. Mountains, farming and conservative values, which he saw as inherited from the peasant revolutionaries who fought for freedom against the elites and oppressors in the Middle Ages, continued to shape the villagers' 'sane' response to a world of increasing disorder. Moritz presented this reaction as one against his inclusion in a global order that was as unavoidable as involuntary.

If Moritz saw exchanges between the valley and the rest of the world as vital, they too often translated into abuse, because outsiders from the urban lowlands and abroad profited from his village and transformed it by engaging in speculation and construction. Incursions of migrants, refugees, and urban dwellers, as well as tourists, the state and the environmentalists with increasing power to influence and decide upon the village's fate 'without having any idea about what it means to live here', were all symptoms of a profoundly worrying situation. This 'takeover' by the 'outside' world was also embodied by Swiss tourism actors whose interest laid in a globalist agenda instead of the village's best interests. Locals, but also outsiders, regularly stated that such processes were abnormal and resulted from locals 'selling their *Heimat*' to greedy outsiders, both materially and symbolically. Contrasting with this reality, Moritz went as far as to imagine his village as a place cut off from the rest of the world. If one were to erect a 'tank barrier' (*Panzersperre*) in the valley, he said, the village and its inhabitants would be on their own – that is how special he felt his village was.

This vision of enforced immobility and militarization felt particularly dystopian to me, but Moritz seemed to evoke it for its appeasing character. In his mouth, this image sounded like a promise of a return to what he described as a genuine state of independence for the 'natives' as 'those who live here'. Such enforced imaginaries of immobility and isolation implied a vision of 'order' that seemed particularly pleasing to other locals, too. If the valley's cosmopolitanization enabled Kirsti and many villagers to express their cosmopolitan sensibilities, the same reality motivated much of the antimobile, antimodern and anticosmopolitan discourses for other dwellers, who treated the influx of outsiders with a general feeling of scepticism.

While the valley's village spirit (*Dorfgeist*) was getting lost in a connected world of flattening differences, the threat was also posed in more cultural terms. As another *Einheimische* I visited on his farm explained, the 'new blood' (*neues Blut*) coming to the valley, the 'outsiders' (*Auswärtige*) and 'foreigners' (*Ausländer*), were particularly worrying. The latter spoke foreign languages that he believed contributed to the disappearance of the typical local Swiss-German dialect that he and the other genuine natives spoke. He was worried that the life and customs of those who, like him, '*really* lived here' were about to die out. 'Real villagers are disappearing', he said, before adding: 'The [Portuguese] kids in front of the school, they speak languages I don't understand. I know *you* are used to it but *I* am having difficulties with it', he said to me and his life partner, who had both grown up and worked in the urban lowlands (*Unterland*). The new blood that had increasingly mixed with the native local population was sometimes described as 'healthy' by some locals, but fundamental differences remained between 'genuine' inhabitants of the Alps and the others.

It is difficult to evaluate precisely the importance of such reactionary attitudes in the Bernese valley and in the Alps generally. A handful of 'natives' I talked to presented visions that were indeed more akin to cosmopolitan ideals of openness towards other cultures. Alex, a young man in his twenties, was one of them: in spite of being very attached to 'traditions' as an apprentice in farming, he expressed gratefulness that so many people from the outside visited his village. Alex had forged meaningful relationships with lowlanders who came to the valley, but he also admitted being somewhat of an exception, and of feeling like an outcast in his own community. When he had a brief relationship with a Black tourist from Berne, his peers mocked him, he remembered. Alex expressed generally feeling uncomfortable when other villagers engaged their nativist, racist rants about the globalized world. Being in a relationship with such an outsider was going against what one would expect of a native. This was confirmed by Adrian, who was one of the only *Einheimische* active in the very multicultural hospitality industry: he too had developed a tolerant attitude and openness for people from other backgrounds, which stood out when he spent time with his native peers. These examples illustrate that 'natives', like any other category of 'host population', cannot be apprehended as a monochromatic picture (Stronza 2005). But Alex and Adrian's accounts, as well as the narratives shared among many of the native people I met, point to the dominance of anticosmopolitan sensibilities in the valley.

What is more, calls to maintain native farming identity and traditions in the valley against the evils of globalization were also shaped by a particular political background. At the time of my first fieldwork between 2017 and 2018, the radical right-wing SVP (the Swiss People's Party) was the most represented party in the region, with more than half of the communal council[11] led by its members. The only alternative to the SVP at the time was the right-wing FDP, with whom the former was closely associated.[12] All the referenda initiated by the SVP in the more recent years found a large support in the municipality, as in the whole Bernese Highlands. To a certain extent, the strong presence of the SVP mirrored the nationwide successes of this party since its radicalization in the 1990s. Since then, the party has adopted a more racist, xenophobic tone, based on an anti-Islam, anti-immigration and anti-EU rhetoric that has deeply influenced national politics. For many analysts of the SVP, it is this antiglobalist, anticosmopolitan narrative – albeit combined with a neoliberal agenda – that has contributed to its rise,[13] giving it access to the highest institutions of the country such as the Federal Council (Skenderovic 2007, 2009). At the same time, the party has also drawn on a long history in Switzerland, and particularly in the canton of Berne (Kriesi et al. 2005). Founded as a farmers' party in the 1930s in the rural Bernese region, early on the SVP found much support in Alpine valleys like my field site.

For Moritz and many of the locals I met, supporting the SVP was described as an 'obvious' choice, strongly anchored in local history and part of a commitment to defend farming with a strong lobby in national politics (as opposed to the FDP Liberals). The left was largely absent from this region: the socialist 'reds' aimed to defend foreigners and 'naive' imaginaries of solidarity while the greens had been involved in romanticized imaginaries of nature (see Chapter 2). By contrast, many of my interlocutors admitted feeling close to the Swiss farming and Alpine imaginary that the SVP regularly mobilized in political campaigns. When speaking of local politics with village dwellers, they would usually explain that the SVP's strong representation in the valley was also a historical fact as the village had always been conservative. As a young *Einheimische* said to me, reactionary politics were simply 'normal' for a rural mountain village, whose problems were not those of the multicultural working class in the lowlands. Such visions were not solely shared by the so-called natives of the village but also reiterated by anticosmopolitan tourists and outsiders, like a couple from a nearby Swiss city, who, after years spent in the village as guests, recently moved to the village for good. They told me that they came to the valley precisely because of its political conservatism and its 'pure' imaginaries, in contrast to the alienating lowlands they came from, which were increasingly inhabited by African migrants.

A majority of the population in the village municipality who were entitled to vote[14] thus largely supported the SVP's controversial, far-right initiatives when put to the national vote. They followed the party's lines since the 1990s, rejecting entry into and even negotiations with the EU. More recently they largely supported its initiatives to limit foreign immigration (the so-called 'initiative against mass immigration' in 2014), thus strengthening 'national preference', and to facilitate the deportation of foreign criminals in 2016. Although many of my interlocutors were sceptical of politics in general, they largely supported the SVP's agenda and values. For Moritz, the radical measures presented by the SVP were necessary to defend 'safe and prosperous' Switzerland, which was being used and abused by foreign criminals who were 'piling up' in prisons.

As the SVP's initiatives found such a large echo nationally, this local support might seem anecdotal. Yet, I suggest that the party's critical accounts of the cosmopolitanization of society found an important traction in my field site precisely because it rested on a long-standing local farming history, *and* could draw on processes of vertical globalization, which, according to the modernist account, only seemed beneficial to the village. Although it would be tempting to focus solely on local party politics and the long-standing success of SVP before and after its radicalization, my aim here is different. The people I met were, for the most part, not involved in the SVP and openly disliked 'party politics' because, they said, the latter were not grounded in 'everyday problems' and matters. 'We vote for people not parties' was a narrative very often asserted by locals. Instead, villagers justified their support for the SVP in the name of Alpine every day concerns and 'common sense' (Lems 2022), shaping this region as a traditional society, closer to nature and 'saner' than the country's lowlands.[15]

Speaking of conservative or even reactionary politics in an Alpine region historically depicted as 'delayed' (Kilani 1984; Niederer, Bätzing and Anderegg 1993) may itself evoke a modernist narrative. Because of their marginalization, rural regions tend to be depicted as conservative and 'backward', as opposed to what constitutes the forward-looking, urbanized 'centre'. My field site's long history of global connection and the mobility of its dwellers radically contradict this interpretation, like most other Alpine villages (Viazzo 1989). Yet, Moritz's yearning for a return to a greater order for his village and country, shared by many locals and nationals, urged me to think about the political inclination of people who inhabit these regions without falling into modernist dichotomies. My encounters with locals and the political make-up of the region highlighted the importance of approaching the cosmopolitanization of the Alps in relation to antimodern sentiments such as those Moritz mobilized. If such conservative sensibilities have been studied in other Alpine regions (Stacul 2003; Lems 2023), they have practically remained unexplored in scholarly accounts on the Swiss Alps.[16] In following these worldviews and attempting to grasp their origins, I found myself going back to some of the most famous critics of modernity: the Romantics. While they undoubtedly shaped the advent of modern tourism, they also influenced much of the everyday discourses of locals and their feelings of unease with the state of their village and homelands – a theme that will also feature in the next chapters of this book.

ROMANTIC (RE)SENTIMENTS AND ALIENATION

In *Integral Europe: Fast-Capitalism, Multiculturalism, Neofascism*, Douglas Holmes follows leaders of Europe's far-right parties and their 'alternative depiction of society', predicated upon a return to 'culture' and 'tradition' (Holmes 2000: ix). From a contemporary context, Holmes excavates the influence of European Romanticism in shaping a form of all-encompassing, 'integralist' sensibility until the present.[17] Clashing with modernity's values of universalism, rationality and individualism, German Romantic author Johann Gottfried Herder, a leading figure of the eighteenth century, stressed the value of a human's 'inner truths and inner ideals', of belonging and rootedness in their community or culture (Berlin 1976: 153; Holmes 2000: 6–7). Herder believed in 'the incommensurability of the values of different cultures and societies' and in their 'incompatibility' (ibid.). Radically questioning the progressive, cosmopolitan embrace of diversity, Herder defined it as absurd, unnatural and disorientating. Following Romantics like Herder, modernity became closely associated with the phenomenon of alienation or estrangement in the German existential philosophical tradition of thought, as a form of uprooting that is not physical but metaphorical, a form of 'relation of relationlessness' to the world (Jaeggi 2014). In a world of increasing cosmopolitanization and 'involuntary' inclusion in a global order of blurring boundaries, such visions of a world in decay have influenced the contemporary, reactionary responses of those who feel discontent and alien in their own land.

From the Alpine ridges historically depicted as healing alienated tourists in the traces of the Romantics, Moritz and his peers did not equate globalization or tourism development with progress but rather with such feelings of alienation, of meaningless and horizonlessness (see also Beiner 2018). This sentiment was shared by many of my interlocutors, who, while working in an industry built on cosmopolitan ideals of embracing diverse cultures, found their inclusion in such a hypermobile and hyperconnected world deeply unsettling – a sign, as Moritz put it, that 'something was going in the wrong direction'. The romantic images that tourism lobbyists displayed now corresponded less and less with the reality one could witness in the village. Moritz shrugged: 'they [the tourism lobbyists and hoteliers] convey beautiful images, but they no longer *have* beautiful images'.

Many mountain dwellers expressed similar feelings. They felt that they were left with empty tourism imaginaries mobilized by powerful actors and various regulators to attract guests or preserve the local environment, as a national and transnational heritage site (see Chapter 3). This froze locals in Disneyland-like landscapes, where innovation was no longer possible. Such feeling is common among those living in tourism destinations, who often feel objectified by what John Urry has termed the 'tourist gaze' (Urry and Larsen 2011). This concept describes the tourists' particular way of perceiving and consuming destinations, landscapes and people while travelling, often expecting them to fit with pregiven imaginaries, such as those of Alpine idylls inspired, in this case, by the Romantics (see Leite 2017). Whereas this critique is often addressed to tourists, Moritz targeted the makers of tourism, whom he believed did not express a genuine concern to preserve and valorize local traditions, identity and heritage. Rather, they were guided by financial interests and privileged urbanites' yearning for a pristine nature they no longer had connection with in the everyday. The village– its farming past and its peculiarity – were celebrated aesthetically, but the essence of these values was getting lost in a world of connection, of flattening of differences, and of speed, as we will see in the next chapter. According to Holmes, these features make for a context of 'fast capitalism', which is fertile ground for the emergence of 'integral politics' because it erodes 'social meaning' (Holmes 2000: 10).[18]

Another symptom of the modern incursions of migrants, lobbyists and cosmopolitans in the valley was what many described as the waning status of the *Einheimische* and its privileges. Because they were 'from here', 'born here' and were 'living here', many natives felt they should be given 'primacy' or 'priority' in local, but also national, politics. Moritz, for instance, stated that the increasing flow of people, material and capital now needed to be regulated. He believed that priority needed to be given to native locals like him instead of outsiders and foreigners. 'I say, we must be first, we were born here, so we must see that our people have work first. Afterwards, only, the others!', Moritz explained. This formulation of nativism, arguing for a 'preference for the "native" exclusively on the grounds of "*being* native"' (De Genova 2016: 233, italics in original), based on an ontological difference between *Einheimische* and non-*Einheimische* people, echoes Herder's interpretation of cultural incommensurability. This also found expression in the concerns over the valley's loss of its core, authentic population, which coloured many of the discussions I had with locals over the impacts of gentrification. The (mostly) native locals I talked to referred to the emigration of native youth – pushed out of their *Heimat*, to more urban areas – as being particularly upsetting and abnormal.[19] When they left, these natives became '*heimweh Einheimische*' (homesick natives), whose connection to the valley, to their origins, would perdure no matter what. The following message, posted by a man on a Facebook group before criticizing the state of a valley that was forever 'his', for instance, poignantly expressed the natural link tying him to his *Heimat*: 'I was born and raised in the village. I am and always will be a villager. My father and grandfather came from the village, and I am now a highlander who moved 'down' (*heruntergekommener Oberländer*). Actually, I should have the right to vote in communal matters like the Swiss abroad. Who are the natives (*Einheimische*) today?'

In the name of such indefectible ties to their valley as natives, superfluous and absurd administrative rules – like this post suggests with the voting regulations – were sometimes transgressed by native locals. One of my interlocutors secretly renovated and expanded his house at night, avoiding the long and strenuous process of asking for official permission at the municipality – an 'absurdity'. A family I met also went to great lengths to try to get their son, who had moved away from the valley and 'down', a local authorization to drive on the valley's special, barred roads even though he was no longer an inhabitant of the municipality and thus no longer eligible for this permit. They did so with the belief that the family's Alpine hut (*Vorsass*), which he would be able to drive to with the authorization, was proof of his genuine belonging that laws and regulations did not formally recognize. Many natives I met claimed the need to return to greater order via more restrictive rules for the outsiders and foreigners degrading the valley, while stating their need, as naturally belonging natives, for greater autonomy and freedom from these rules (see Chapter 4).

THE NATIVE IN HIS ALPINE *HEIMAT*: LOCAL AND NATIONAL IMPLICATIONS

When utilized as a contemporary diagnostic, alienation and its accompanying feelings of deracination, meaninglessness, horizonlessness or relationlessness all lament a negative state that needs to be understood against ideals of a good, meaningful, fulfilling life. In the German-speaking and particularly the Alpine context, the idea of the *Heimat*, as the place where one naturally belongs, shapes this background (Cole and Wolf 1999). The feeling of *Heimat*, of deep attachment to valley, was manifold and indescribable, many locals contended. It comprised of personal memories, of childhood, of a house, a barn or family, often all at once. This mirrored the *Heimat*'s all-encompassing nature, with material, affective, aesthetic considerations pertaining to one's house, one's birthplace or one's nation (Leimgruber 2019: 143–44). Feeling grounded in one's *Heimat* was a 'privilege', said Franziska, a native local I will introduce in more detail later on. Living as a native in one's

Heimat represented a normative, universal horizon that everyone should pursue, and those who did not lived in a state of false consciousness, she implied. The *Einheimische* – literally those at home, in their *Heimat* – clearly epitomized such natural, harmonious being 'at home' in the valley.

Confronted with the important circulation of clichés and stereotypes about the *Einheimische* in the village and in Switzerland, I could paint a relatively precise picture of Moritz, before even meeting him. His farming ancestors, his typical local family name, his inherited house on the scattered settlement towards the outskirts of the village, the Alpine hut (*Vorsass*) he owned and the barred roads he could drive along thanks to the sticker on his 4×4, as well as his local Swiss-German dialect, were all features I had come to expect from natives. Besides having the village municipality inscribed as one's legal *Heimatort* – place of origin – on one's identity card, enjoying the *Einheimische* status also implied a particular type of family history, as well as social, economic and material relations that, as we will see, were difficult to obtain for outsiders, be they from the next village or from another country. Moritz drew on all those markers of natives and also fitted the gendered, masculine implications of this label. As locals told me, in a version of Alpine virilocality, traditionally women had tended to get married and to emigrate more permanently, contrary to men, who often stayed as heads of farms and property.[20] Such more or less institutionalized features of genuine rootedness in the village did not always align neatly; they were subjected to changes and reinterpretations as people moved in and out of the village, but remained central in my interlocutors' understanding of the valley and its population.

When I was encouraged to speak to the *Einheimische* by many villagers, I was also told to expect nothing less than coldness and distance from the genuine natives, whose attitude I should read as both 'backwardness' and as a commitment to 'freedom'. Countless locals explained that the surrounding mountains determined the native's close-mindedness, stubbornness and reactionary mindset, because walls of rock and ice blocked their horizons, literally and figuratively, or cast giant shadows over the valley in winter. In doing so, they mobilized a common narrative of environmental determinism in the Alps, as various authors on the region have shown (Stacul 2003; Dickinson 2010).[21] As Moritz himself expressed, the Alpine topography of the valley profoundly inspired the definition of the real native's features as autochthons. Natives were, following the Greek origins of the term 'autochthony', 'born from the soil' (Geschiere 2009, 2011). And while the category of the native seemed pure, superior most of the time, it was also associated with the narrow-minded nature of natives and a ground to ridicule (some of) them, as I will discuss in the next chapters. As such, the native illustrated what Peter Geschiere has described of the ambiguous nature of figure of the autochthon, seen both as 'prestigious first come but also as primitive or even prehuman' (Geschiere 2009: 11).

The native villager was a figure who was embedded in discourses about local pride of the village, as a peculiar place in a world of increasing homogeneity. This was evidenced by sentences taken from the local anthem stating 'there is no valley more beautiful than [this one]', regularly used in political meetings, on websites of local associations and on cowbells displayed during farmers' events. The presence of tourists from all over the world also stood as proof of the valley's difference and superior quality and, by association, that of its dwellers, as a local woman once told me after hearing how a tourist seated nearby had expressed heartfelt emotion at the sight of the mountains. However, such a localist version of autochthony also had greater implications that Moritz revealed when he spoke of the admirable nature of the peasants he was descended from as a native villager.

Moritz insisted that events opposing the 'peasants with halberd' to 'foreign bailiffs' in the Middle Ages – historically recorded events[22] – were 'as if taken from the novel' (*wie aus dem Roman*). As such, they fed into a romanticized tale that he and other locals narrated, highlighting the proximity between their ancestors and those who were said to be the founders of Switzerland, like the national hero Wilhelm Tell. As the myth has it, Tell rebelled and assassinated the tyrannical Habsburg bailiff named Gessler sometime in the fourteenth century. Over time, Tell came to epitomize the origins of the Swiss Confederation and the Old Swiss Confederates[23] – the *Eidgenosse* – and their foundational struggle for freedom against the imperial powers of the Habsburgs (Zimmer 1998: 648). When Friedrich Schiller depicted Tell's story in the Alps, he did so to utter a critique against a contemporary decaying civilization, linking Tell with the myth of the morally superior, honest Swiss Alpine 'peasants and shepherds' (Bendix 1992; de Capitani 2010). This association soon became dominant everywhere in the country and beyond.

Multiple *Einheimische* interlocutors of mine explicitly referred to the values of the Old Confederate – the *Eidgenosse* – to which they felt particularly close because they too lived in the mountains that instilled ideals of freedom, far away from foreign imperialism (see Chapter 5). Historian Rubi historicized the farming past as well when he stressed the exceptional, superior, free identity of mountain dwellers as those who had led the farmer revolts in the region as 'outstanding figures', 'free and unafraid' (Rubi 1997: 18). Highlanders, he stated, were 'a breed of their own', of a certain 'insolence' (*Auflüpfigkeit*) in spite of the cosmopolitan feel of their valleys (ibid.). Evoking the farming history and Alpine geography enabled Moritz, Rubi and others to perform the highlanders' native status and genuine belonging to the Alps but also the country.

This is hardly surprising, given that the Alps have played a pivotal role in the Swiss nation-building process, shaping the very definition of the 'original Swiss', the *Eidgenosse* or even *Homo Alpinicus Helveticus* (Marchal and Mattioli 1992; Zimmer 1998, 2003).[24] At the same time, Celia Applegate's work on the concept of *Heimat* (1990) discusses the importance of such regionalisms – in this case Alpine – as part of broader visions of the nation, prominent in German-speaking areas of the world. In this context, she illustrates how a commitment to local or regional identity becomes a distinctive way of mobilizing everyday forms of national belonging – especially when one hails from the mythical Swiss Alps. Although direct references to Tell were rare during my fieldwork, the national imagery of Alpine identity, farming roots and freedom made Moritz and his peers into Tell-like figures, embodying Switzerland's origins as an Alpine nation of stubborn, freedom-loving peasants, resisting elites and cosmopolitans from the lowlands. By contrast, as a Swiss from the French-speaking part and from the more urbanized, left-leaning lowlands, my genuine belonging in an original ideal of Switzerland was regularly undermined. While such narratives often led on to an essentialization of *the* native identity, the next chapters will reveal the strong tensions and conflicts among the native population of the valley.

LINES OF UN-/BELONGING

When I met Kristi and her friends on various occasions, they told me about their attachment to village's environment, its urban flair, the friendships they had made, the families they had built and the fulfilling work–life balance they had cultivated in it. As I got to know them better, they often introduced me to a more contested story of belonging. If they personally considered the village as their *Heimat*, they explained how they felt excluded from normative definitions of belonging in the valley. One evening, as I was sitting in a local restaurant with some of Kristi's friends, Simone, who had lived in the valley for most of her life, nodded in the direction of a neighbouring table. Simone guessed that these customers were probably a typical *Einheimische* family, judging by their dialect and the men's clothing style: farmers' clothes and hats displaying local associations. One of her friends responded that Simone herself was probably a native too, given the years she had been living in the village. 'I doubt I would even really become one', Simone said with a shrug. Sure, she was an *Einheimische* on paper, as a legal resident of the valley. She had access to *Einheimische* (in this sense, meaning 'local') prices when buying a ski pass after presenting her residence certificate or when going for lunch in the restaurant next to her workplace. She also considered the valley her *Heimat*, especially now that her sister's family had also come to live in the valley after spending their lives in the lowlands. She also pictured herself having a family here. Yet, in spite of such legal and affective marks of belonging to the village, she would never call herself an *Einheimische* – 'in her *Heimat*' – as the *Einheimische* themselves made her feel like she did not genuinely belong. With her lowlander dialect from the area of Zurich, her untypical family name drawing on her 'outsider' origins, she doubted that she would ever be considered a local, unless she were to marry a native man, as many Swiss women did.

Simone was upset by this situation. Not only was she not considered a genuine insider in a village she had been living in for years, but native locals remained 'distant' and 'cold' to her. She had heard explanations claiming that the coming and going of tourists and seasonal workers had further legitimized the *Einheimische*'s cold mountain people character. One native villager told her it was not 'worth' building relationships with such fleeting crowds – something I had been told, too. But for Simone, these explanations did not make sense. She was firmly rooted in the village and involved in many of the local clubs and activities, not to mention that she too contributed to the economic development of the place. Another friend at the table noted that the distance separating them from the *Einheimische* locals was also a more practical one, as many of the latter rarely ventured into the village centre anyway. Simone's friend explained that these villagers complained when they had to come 'down' to the centre of the village, to the crowded *Hauptstrasse*, to fetch provisions and orders in the shop where she worked.

Similar to what Yuval-Davis (2006) describes, the forms of belonging I encountered in the valley were elusive, personal or 'difficult to describe with words', as someone once put it, as well as materially and socially anchored in the valley's political structure. There were, according to Simone, Kirsti and many of their friends, boundaries traced by others that made them feel like outsiders in a place they considered home. Their experiences reveal how calls for a return to greater order – to a rooted, authentic *Heimat* or farming values, as well as laments on the disappearance of local identity in the international resort – did not operate in a vacuum. They were part of the everyday social work that was put into (re)creating and maintaining boundaries and categories of belonging 'despite a flow of personnel across them' (Barth 1998: 9; see also Lamont and Molnár 2002). Against the claim made by many villagers that nativeness had lost traction as a meaningful category following globalization (for good or bad), experiences of being *made* not to belong for locals who were as culturally proximate as Simone (as a German-speaking Swiss who had spent years in the valley) reveal the ongoing effects of nativism as a dominant political and material structure in the valley.

Like Simone, many of the 'outsiders' (*Auswärtige*) or 'newcomers'(*Zugezogene*) in the village recounted the difficulty, if not impossibility, of becoming a native in one's lifetime, even for those who came from places located closer than Simone and Kristi's urban hometowns. One man from the Bernese Highlands who had lived in the village for the past thirty years and married an native woman,[25] for instance, confirmed that based on his observations, it took three generations to become native. According to him and Simone, and echoing the views of Yuval-Davis, gender also played a role in his exclusion from local definitions of nativeness. Had he been a woman, he believed that he would have been integrated more seamlessly as a spouse (*Zugeheirateter*) in the valley because of the traditional system of kinship in the Alps, which had historically favoured the immigration of women from outside of the valley's confines (Heady 1999).

Various cases of women married to native men nonetheless contradicted this widespread narrative in the valley. As a woman from the neighbouring valley put it, she was still treated like an outsider by the genuine, native locals, as a result of having been 'imported here, even if not from very far away'. Two other older men whose parents were born in the village explained that they too were not considered *Ureinheimische* because of their non-native family names. This was the case for many hoteliers too, who also overwhelmingly drew on (often temporally distant) outsider origins. In spite of their dialect, their involvement in the village life and a local family history anchored in the village, they were still not perceived as genuine autochthons. Although Simone and her friends often regretted the distance created between native insiders and themselves, others, like these men, had accepted their fate as part of the reality of living in this village.

The discussions I had with Swiss locals about belonging often involved misunderstandings, especially at the beginning of my research. When using the word '*Einheimische*', my interlocutors sometimes switched between a more nativist understanding of it – meaning the genuine, original native or autochthon – and a more descriptive or wider-encompassing understanding of the term – meaning local – which I initially found difficult to differentiate. As my fieldwork progressed, the omnipresence of the normative, nativist status of the *Einheimische* became clear, thanks to my exchanges with people like Moritz and Simone. Yet, one category of inhabitants in the valley escaped such (relative) imprecision: the so-called *Ausländer*, or foreigners.

A foreign *Einheimische* was a contradiction in terms: 'a foreigner cannot be a native', expressed one of my local native informants. When I asked for more detail on this, he explained that these were just how things were. As I then heard from fellow natives, locals and migrants themselves, an accent, a cultural background, a lack of interest in 'integrating' with the local and national community, were seen as unsurmountable differences hindering any form of organic integration. While the capacity of Swiss outsiders to ultimately be referred to as native was limited in practice, as Simone's example shows, many villagers flat-out denied the possibility that foreigners could 'genuinely' ever belong. Migrants residing in the village were excluded in both localist and nationalist variations of nativism. If the valley was made up of the simultaneous presence of people from different backgrounds, there were clear divides and hierarchies among them: on the one side of the spectrum of belonging, the *Einheimische*, and on the other, the foreign migrants.

CONCLUSION

Starting from the radically different visions of the village proposed by Kirsti and Morris, in this chapter I have described the single reality they belong to: that of a globalized, yet profoundly conservative, resort. While international tourism often evokes progressive, cosmopolitan imaginaries of exchange in tourism promotion, everyday discourses and academic literature, the case presented here reveals that this relationship cannot be seen as an obvious or natural one – on the contrary. Anticosmopolitan or nativist views sprout from this same globalized context, inspired and nurtured by some of the same Romantic critical discourses of modern institutions that have shaped the advent of tourism. While cosmopolitan sensibilities find an echo in this touristic, transnationally connected valley, reactionary worldviews also fixate on the (threatened) ideals of authenticity and purity that the Alps and its inhabitants represent. Although I will further explore how Romantic worldviews play a particular role in an Alpine touristic economy in this regard in chapters 2 and 3, my field site is emblematic of the trend that many anthropologists have examined over recent years: the essential coexistence of a transnationally connected world *and* right-wing populism or nativism that rejects, or aims to reform or reorder, the modern institutions that have often contributed to people's (relative) well-being (Mazzarella 2019; Pasieka 2019).

What emerges through the experiences of Kirsti and Moritz I have discussed here is not simply the coexistence of their worldviews in the same reality, but also the historical and political dominance of one over the other. In spite of what Kirsti described, there was little visible trace of 'multiculturalism' in the political sense of the word in the valley.[26] Instead of a political will to cherish and foster a co-presence of and exchange between various cultures, anticosmopolitanism and nativism structured the valley's politics of belonging (Leitenberg 2025). On both local and (trans-)national level, they celebrated and privileged the primordial figure of the *Einheimische* against other locals, whether Swiss outsiders or foreign migrants. Politically, they positioned a rural Alpine vision in opposition to a leftist, progressive, and urban model. This in no way means that *all* villagers of all backgrounds shared this vision, but it was a hegemonic vision. Far from disappearing with the advent of capitalist tourism, anticosmopolitanism has developed alongside comsopolitanization, sometimes feeding off its discontents.

Having outlined the general context shaping my field site, as well as the exclusionary dynamics at stake when speaking of 'natives', I will now, in the following chapters, turn to events that have shaped debates around tourism dependency, especially with regards to native life and its superior character. In the last two chapters of the book, I will explicitly focus on those excluded from such native life.

NOTES

1. This narrative was presented on the website of the municipality until recently.
2. Historian Bernard Schär notes that the peasants and shepherds were celebrated as morally superior, European 'good savages', exotic but close, as opposed to the 'bad' savages in the faraway tropics (Schär 2012, 2015).
3. Owing its name to the Grand Tours that were going through Europe in the nineteenth century.
4. See Friedli 1905; Kröner 1968; Nägeli-Oertle 1986; Rubi 1986; Wiesmann 1986a, 1986b; Messerli 1989; Tiefenbach 2013; Balmer 2015.
5. The distances and differences between the scattered settlements and the village centre (the village street) were greater at that time. There were also different schools gathering pupils of the scattered settlement according to their belonging to multiple areas of the valley.
6. See for example Wiesmann (1986b: 45).
7. The percentage of holiday homes was very high in the village (approximately 65 per cent of the municipal housing infrastructure as of 2017) and was a subject of contention. Since 2015, the number of these houses has been limited nationwide thanks to the 2012 Second Home regulation (*Zweitwohnunggesetz*, ZWG). This limitation was a setback for local construction companies, which had to expand their activities.
8. This refugee accommodation was located far away from the centre of the village (see Chapter 5). It closed in 2018.
9. This centre was located closer to the *Hauptstrasse* than the previous one.
10. Cosmopolitanism is a notoriously fuzzy concept. I here rely on Hannerz's definition of it as being as an attitude of openness or 'a willingness to engage with the Other', and a way of cherishing one's 'relationship to a plurality of cultures'(Hannerz 1990: 847).
11. Representing the local executive powers.
12. Since 2020, the communal council has been led by a mayor that is not affiliated with a party, which many of my interlocutors had thought would be 'impossible'. This was interpreted as a result of the close association between tourism lobbyists and the municipal council by many of my interlocutors. See Chapter 3 for a more detailed discussion.

13. The rise of the right-wing populism is often attributed to a frustration with globalization in other Western countries (Gingrich and Banks 2006; Gusterson 2017).
14. Meaning to be Swiss, to be eighteen years of age, and to have been a permanent resident in the village for more than three months.
15. The return to forms of everyday 'common sense' in a world led by cosmopolitan, technocratic elites is a common trope in the rise of right-wing populism. See also Wodak 2015 and Pasieka 2022.
16. One exception could be the work of Petite (2011, 2013) and the contestation of lowlander, left-wing politics. He does not explore Alpine anticosmopolitanism as such, however.
17. He does so by heavily relying on Isaiah Berlin's work on counterenlightenment (Berlin 1976).
18. See chapters 3 and 4 for a more detailed discussion of the inclusion of Romantic critique in local politics.
19. The reasons evoked were, for instance, the consequences of the Second Home law, which was said to have increased the real estate prices in the region, making it difficult for young native families to remain in their village. See also Chapter 5.
20. As Leimgruber describes, the concept of *Heimat* is deeply linked with the notion of material heritage, the property of land and a household, which traditionally has been owned by men (Leimgruber 2019: 145–46). See also Heady 1999.
21. See also Berthoud 1972; Debarbieux 2018; and Heady 1999.
22. Like many other regions of the Bernese Highlands, the locals also fought against the Covent of Interlaken in 1348–49.
23. The Confederates' Oath of the Rütli, which was said to have taken place in 1307, is the foundational event of Swiss national history.
24. Preoccupations concerning the purity of the 'Swiss race' as '*Homo Alpinicus Helveticus*' gave way to 'eugenicist discourses and practices elaborated by internationally renowned Swiss scientists' such as psychiatrist August Forel (1848–1931) and anthropologist Otto Schlaginhaufen (1879–1973) (Michel 2015; Mottier 2000: 413).
25. The only case that I knew of an *Einheimisch* man marrying an outsider woman.
26. I here refer to agendas and political ideals that would favour the coexistence of different cultural groups, and be influenced by a certain cosmopolitan sensibility. 'Mutliculturalism' as a concept has, however, been critiqued because of deriving from the 'nation-state social fabric of cultural differences' (Beck 2011: 54).

REFERENCES

Applegate, Celia. 1990. *A Nation of Provincials: The German Idea of Heimat*. Berkeley: University of California Press.

Balmer, Katharina. 2015. *Als Gletscher Noch Aus Eis Waren: Die Jungfrauregion Einst Und Jetzt*. Bern: Ott-Verlag.

Barth, Fredrik (ed.). 1998. *Ethnic Groups and Boundaries: The Social Organization of Culture Difference*. Long Grove, IL: Waveland Press.

Barton, Susan. 2008. *Healthy Living in the Alps: The Origins of Winter Tourism in Switzerland, 1860–1914*. Manchester: Manchester University Press.

Beck, Ulrich. 2011. 'Multiculturalism or Cosmopolitanism: How Can We Describe and Understand the Diversity of the World?' *Social Sciences in China* 32(4): 52–58.

Beiner, Ronald. 2018. *Dangerous Minds: Nietzsche, Heidegger and the Return of the Far Right*. Philadelphia: University of Pennsylvania Press.

Bendix, Regina. 1992. 'National Sentiment in the Enactment and Discourse of Swiss Political Ritual', *American Ethnologist* 19(4): 768–90.

Bergier, Jean-François, et al. 2013. 'Alpes'. *Dictionnaire Historique De La Suisse*.

Berlin, Isaiah. 1976. *Vico and Herder: Two Studies in the History of Ideas*. London: Hogarth Press.

Berthoud, Gérald. 1972. 'From Peasantry to Capitalism: The Meaning of Ownership in the Swiss Alps', *Anthropological Quarterly* 45(3): 177–95.

Chio, Jenny T. 2014. *A Landscape of Travel: The Work of Tourism in Rural Ethnic China*. Seattle: University of Washington Press.

Cole, John W., and Eric R. Wolf. 1999. *The Hidden Frontier: Ecology and Ethnicity in an Alpine Valley*. Los Angeles: University of California Press.

Debarbieux, Bernard. 2018. 'How Can One Be a "Montagnard"? Social and Political Expressions of Modern Imaginaries of Territoriality', in Christos Kakalis and Emily Goetsch (eds), *Mountains, Mobilities and Movement*. London: Springer, pp. 129–53.

Debarbieux, Bernard, Cristina Del Biaggio and Mathieu Petite. 2008. 'Spatialités et territorialités du tourisme. Dialectique du flux et de l'ancrage dans Les Alpes', *Civilizations. Revue Internationale d'Anthropologie et de Sciences Humaines* (57): 75–89.

de Capitani, François. 2010. 'Peuple des Bergers', *Dictionaire Historique De La Suisse*.

De Genova, Nicholas. 2016. 'The "Native's Point of View" in the Anthropology of Migration', *Anthropological Theory* 16(2–3): 227–40.

Denning, Andrew. 2014. *Skiing into Modernity: A Cultural and Environmental History*, Vol. 3. Oakland: University of California Press.

Dickinson, Edward. 2010. 'Altitude and Whiteness: Germanizing the Alps and Alpinizing the Germans, 1875–1935', *German Studies Review*: 579–602.

Enzensberger, Hans Magnus. 1996. 'A Theory of Tourism', *New German Critique* (68): 117–35.

Ferguson, James. 1999. *Expectations of Modernity: Myths and Meanings of Urban Life on the Zambian Copperbelt*, Vol. 57. Los Angeles: University of California Press.

Friedli, Emanuel. 1905. *Bärndütsch Als Spiegel Bernischen Volkstums: Alphabetischer Nachweiser Zu Band Saanen*. Bern: Verlag Von a. Francke.

Geschiere, Peter. 2009. *The Perils of Belonging: Autochthony, Citizenship and Exclusion in Africa and Europe*. Chicago: University of Chicago Press.

———. 2011. 'Autochthony, Citizenship and Exclusion: New Patterns in the Politics of Belonging in Africa and Europe', in Sarah Albiez et al. (eds), *Ethnicity, Citizenship and Belonging: Practices, Theory and Spatial Dimensions = Etnicidad, ciudadanía y pertenencia: prácticas, teoría y dimensiones espaciales*. Madrid: Iberoamericana, 175–98.

Gingrich, André, and Marcus Banks. 2006. *Neo-Nationalism in Europe and Beyond: Perspectives from Social Anthropology*. New York: Berghahn Books.

Gusterson, Hugh. 2017. 'From Brexit to Trump: Anthropology and the Rise of Nationalist Populism', *American Ethnologist* 44(2): 209–14.

Haller, Tobias, et al. 2019. *The Commons in a Glocal World: Global Connections and Local Responses*. London: Routledge.

Hannerz, Ulf. 1990. 'Cosmopolitans and Locals in World Culture', *Theory, Culture & Society* 7(2–3): 237–51.

Heady, Patrick. 1999. *The Hard People: Rivalry, Sympathy and Social Structure in an Alpine Valley*. Abingdon: Routledge.

Holmes, Douglas R. 2000. *Integral Europe: Fast-Capitalism, Multiculturalism, Neofascism*. Princeton, NJ: Princeton University Press.

Jaeggi, Rahel. 2014. *Alienation*. New York: Columbia University Press.

Kilani, Mondher. 1984. 'Les images de la montagne au passé et au présent. L'exemple Des Alpes Valaisannes', *Schweizerisches Archiv Für Volkskunde Basel* 80(1–2): 27–55.

Kriesi, Hanspeter, et al. 2005. *Der Aufstieg der SVP. Acht Kantone im Vergleich*. Basel: NZZ-Verlag.

Kröner, Arlinde. 1968. *Grindelwald: Die Entwicklung eines Bergbauerndorfes zu einem internationalen Touristenzentrum: Ein Beitrag zum Problem des kulturgeographischen Wandels alpiner Siedlungen*. Vol. 74, *Stuttgarter Geographische Studien*. Stuttgart: Im Selbstverlag Des Geographischen Instituts Der Universitaet Stuttgart.

Kuhn, Konrad. 2019. 'Unsichtbare Dienste und schmutzige Wäsche: Zu Arbeitswelt und Lebenssituation von Zimmermädchen', in Andrea Leonardi, Timo Heimerdinger and Evelyn Reso (eds), *Hotelpersonal – Lebens- und Arbeitsalltag im Dienste des Tourismus*. Innsbruck: Studien-Verlag, pp. 295–313.

Lamont, Michèle, and Virág Molnár. 2002. 'The Study of Boundaries in the Social Sciences', *Annual Review of Sociology* 28(1): 167–95.

Leimgruber, Walter. 2019. 'Constructing a Home: Heimat as Expression of Privilege, Belonging, Exclusion and Identity', *Journal for European Ethnology and Cultural Analysis* 1: 142–67.

Leite, Naomi. 2017. *Unorthodox Kin: Portuguese Marranos and the Global Search for Belonging*. Oakland: University of California Press.

Lems, Annika. 2016. 'Ambiguous Longings: Nostalgia as the Interplay among Self, Time and World', *Critique of Anthropology* 36(4): 419–38.

———. 2022. 'Deciphering Everyday Meaning-Making with Gramsci', *Dialectical Anthropology* 46: 395–415. Doi: 10.1007/S10624-022-09658-5.

———. 2023. 'Anti-Mobile Placemaking in a Mobile World: Rethinking the Entanglements of Place, Im/Mobility and Belonging', *Mobilities* 18(4): 620–34. Doi: 10.1080/17450101.2023.2220942.

Marchal, Guy P., and Aram Mattioli. 1992. *Erfundene Schweiz: Konstruktionen nationaler Identität*. Zurich: Chronos.

Mathieu, Jon, and Simona Boscani Leoni. 2005. *Die Alpen! Zur europäischen Wahrnehmungsgeschichte seit der Renaissance = Les Alpes! Pour une histoire de la perception Européenne depuis la Renaissance* 1. Aufl. Edn. Vol. 2, *Studies on Alpine History*. Bern: P. Lang.

Mazzarella, William. 2019. 'The Anthropology of Populism: Beyond the Liberal Settlement', *Annual Review of Anthropology* 48(1): 45–60.

Messerli, Paul. 1989. *Mensch und Natur im apinen Lebensraum: Risiken, Chancen, Perspektiven: Zentrale Erkenntnisse aus dem Schweizerischen MAB-Programm*. Bern: P. Haupt.

Michel, Noémi. 2015. 'Sheepology: The Postcolonial Politics of Raceless Racism in Switzerland', *Postcolonial Studies* 18(4): 410–26.

Moser, Patrick. 1997. *'So Wird die Jungfrau zur Demoiselle Gemacht': Projektierung und Bau der Jungfraubahn*. Zurich: Chronos.

Mottier, Véronique. 2000. 'Narratives of National Identity: Sexuality, Race and the Swiss "Dream of Order"', *Revue Suisse de Sociologie* 26(3): 533–58.

Nägeli-Oertle, Rudolf. 1986. *Die Berglandwirtschaft und Alpwirtschaft in Grindelwald: Schlussbericht zu Projekt 4.184*. Bern: Bundesamt Für Umweltschutz.

Netting, Robert M. 1981. *Balancing on an Alp: Ecological Change and Continuity in a Swiss Mountain Community*. Cambridge: Cambridge University Press.

Niederer, Arnold, Werner Bätzing and Klaus Anderegg. 1993. *Alpine Alltagskultur Zwischen Beharrung und Wandel: Ausgewählte Arbeiten aus den Jahren 1956 bis 1991*. Bern: Paul Haupt.

Nöbauer, Herta. 2021. 'Weather, Agency and Values at Work in a Glacier Ski Resort in Austria', in Paul Sillitoe (ed.), *The Anthropocene of Weather and Climate: Ethnographic Contributions to the Climate Change Debate*. New York: Berghahn Books, pp. 124–45.

Pasieka, Agnieszka. 2019. 'Anthropology of the Far Right: What If We Like the "Unlikeable" Others?' *Anthropology Today* 35(1): 3–6.

———. 2022. 'Theft of Gramsci? On the Radical Right, Radical Left and Common Sense', *Dialectical Anthropology* 46(4): 417–36. Doi: 10.1007/S10624-022-09681-6.

Petite, Mathieu. 2011. *Identités en chantiers dans les Alpes: Des projets qui mobilisent objets, territoires et réseaux*, Vol. 20. Bern: Peter Lang.

———. 2013. 'Les Montagnards Face aux Ecolos. Une Enième Déclinaison au Travers de l'initiative Weber', *Journal of Alpine Research* (Hors-Série).

Rubi, Rudolf. 1986. *Vom Bergbauerndorf zum Fremdenort: Gastgewerbe, Alpinismus*. Vol. 2, *Im Tal von Grindelwald. Bilder aus seiner Geschichte*. Grindelwald: Verlag Sutter Druck AG.

———. 1987. *Der Sommer- und Winterkurort: Strassen Und Bahnen, Wintersport*. Vol. 3, *Im Tal von Grindelwald: Bilder aus seiner Geschichte*. Grindelwald: Verlag Sutter Druck AG.

———. 1993. *Das Gletscherdorf: Von den Grindelwaldgletschern, der Gletscherpfarrer und sein Dorf*. Vol. 5, *Im Tal von Grindelwald. Bilder aus seiner Geschichte*. Grindelwald: Verlag Sutter Druck AG.

———. 1997. 'Ein Streifzug durch die Geschichte Grindelwalds', in Fritz Balmer, Rudolf Rubi and Christian Baumann (eds), *850 Jahre Grindelwald [1146–1996]: Bilder und Notizen zum Jubiläumsjahr*. Grindelwald: Gemeinde Grindelwald, pp. 16–18.

Salazar, Noel B., and Nelson H.H. Graburn. 2014. *Tourism Imaginaries: Anthropological Approaches*. New York: Berghahn Books.

Sampaio, Sofia, Valerio Simoni and Cyril Isnart. 2014. 'Tourism and Transformation: Negotiating Metaphors, Experiencing Change', *Journal of Tourism and Cultural Change* 12(2): 93–101.

Schär, Bernhard. 2012. 'Bauern und Hirten Reconsidered. Umrisse Der «erfundenen» Schweiz im imperialen Raum', in Patricia Purtschert, Barbara Lüthi and Francesca Falk (eds), *Postkoloniale Schweiz. Formen und Folgen eines Kolonialismus ohne Kolonien*. Bielefeld: transcript, pp. 315–31.

———. 2015. 'On the Tropical Origins of the Alps', in Patricia Purtschert (ed.), *Colonial Switzerland*. Palgrave Macmillan: Basingstoke, pp. 29–49.

Schelbert, Leo. 1976. *Einführung in die schweizerische Auswanderungsgeschichte der Neuzeit*. Zurich: Stäubli.

Skenderovic, Damir. 2007. 'Immigration and the Radical Right in Switzerland: Ideology, Discourse and Opportunities', *Patterns of Prejudice* 41(2): 155–76. Doi: 10.1080/00313220701265528.

———. 2009. *The Radical Right in Switzerland: Continuity and Change, 1945–2000.* New York: Berghahn Books.

Stacul, Jaro. 2003. *The Bounded Field: Localism and Local Identity in an Italian Alpine Valley*. Vol. 18. New York: Berghahn Books.

Starobinski, Jean. 1966. 'The Idea of Nostalgia', *Diogenes* 14(54): 81–103.

Stronza, Amanda. 2005. 'Hosts and Hosts: The Anthropological of Community-Based Ecotourism in the Peruvian Amazon', *NAPA Bulletin* 23(1): 170–90.

Tiefenbach, Marianne. 2013. *Alpkorporationen – Traditionelle Institutionen nachhaltiger Landschaftsentwicklung: Das Beispiel der Bergschaften Grindelwalds im Kontext aktueller gesellschaftlicher Herausforderungen.* Grindelwald: Sutter Druck.

Tissot, Laurent. 2006. 'Développement des transports et tourisme: Quelles relations?' *Revue Suisse d'Histoire* 56(1): 31–37.

Urry, John, and Jonas Larsen. 2011. *The Tourist Gaze 3.0.* London: Sage.

Viazzo, Pier Paolo. 1989. *Upland Communities: Environment, Population and Social Structure in the Alps since the Sixteenth Century*, Vol. 8. Cambridge: Cambridge University Press.

Wiener, James Blake. 2022. 'Die Walserzüge', *Blog National Museum*, 22 November 2022. Retrieved 22 April 2024 from https://blog.nationalmuseum.ch/2022/11/Die-Auswanderung-Der-Walser/?Fbclid=Iwar10mvfkfuocy7xcexzeambcaknjzmflr_Druii9-Vh9kvxzogfj0naxzp0%20%28Accessed%2025%2F06%2F20224%29.

Wiesmann, Urs. 1986a. *MAB-Grindelwald: Die Entwicklung des Fremdenverkehrs und dessen wirtschaftliche, gesellschaftliche und räumliche Bedeutung.* Bern.

———. 1986b. *wirtschaftliche, gesellschaftliche und räumliche Bedeutung des Fremdenverkehrs in Grindelwald.* Vol. 24, *Schlussberichte Zum Schweizerischen MAB-Programm.* Bern: Bundesamt Für Umweltschutz Bibliothek.

Wodak, Ruth. 2015. *The Politics of Fear: What Right-Wing Populist Discourses Mean.* London: Sage.

Yuval-Davis, Nira. 2006. 'Belonging and the Politics of Belonging', *Patterns of Prejudice* 40(3): 197–214.

Zanini, Roberta Clara. 2013. 'Dynamics of the Population and Dynamics of the Memory of an Alpine Community on the Border', *Journal of Alpine Research* 101(3).

Zimmer, Oliver. 1998. 'In Search of Natural Identity: Alpine Landscape and the Reconstruction of the Swiss Nation', *Comparative Studies in Society and History* 40(4): 637–65.

———. 2003. *A Contested Nation: History, Memory and Nationalism in Switzerland, 1761–1891.* New York: Cambridge University Press.

CHAPTER

2

ACCELERATING INTO THE FUTURE

December 2020, on the valley floor, 1.5 kilometres below the village's train station, Mr Frei, CEO of the JR company, gets on stage in the newly built terminal of the 'project of the century': his dream cable car project. I am watching the big day from my computer screen during lockdown in my apartment in the Swiss lowlands. Moments later, cameras will capture his first trip on the brand new gondola from the valley floor to the mountain top, twice as fast as used to be the case before the cable car was in place. I expected to see a bigger smile on his face, but his restraint seems appropriate: we are in the middle of one of the countless 'waves' of the Covid-19 pandemic, and so looking forward to anything else would be insensitive, if not naive. Frei's light-khaki trousers, sky-blue shirt and navy jacket contrast with the red-and-white Swiss flag behind him. His voice echoes in the new concrete hall: 'Behind the idea of the new cable car is the idea that standing still (*Stillstand*) and resting on one's laurels is no option', says Frei. 'JR faces great challenges with the unforeseeable Corona pandemic, like it was already the case with the Second World War. But it is precisely this cable car that makes us unbelievably competitive (*konkurrenzfähig*), that puts us on the front line. We are ready for the future and the coming recovery and boom (*Aufschwung*). Today, we are celebrating a historic day thanks to the help and support of all of you!' Following Frei's intervention, various politicians, economic players and journalists praise him in the new terminal. He is a man capable of literally moving mountains through his will – a pioneer, a saviour, they say. 'If only this [Corona] virus were not ruining the party, everyone here would be cheering, hugging and thinking how this project would not have been imaginable in their wildest dreams!' exclaims the federal chancellor, the representative of one of the highest institutions of the country, who has come for the occasion.

More than Frei's personal achievement, the inauguration of the new cable car marked the final point of nearly ten years of debates, struggles, fears and hopes for most of my interlocutors. After so much time spent imagining, representing, evoking the cable car with them, there was something surreal about the fact that the infrastructure now firmly stood in the valley. What had been anticipated for so long now marked 'history', in Frei's words. I could not help thinking of those who had fiercely opposed and rejected this 'generation's project' (*Generationenprojekt*) and would now have to live with what they despised, before it even materially existed. At the same time, the cable car supporters, like the boss himself, must feel worried too, I thought, as the valley, like the world, was moving into uncharted territory. I imagined opponents feeling some sense of relief, perhaps revenge even, in seeing so many of the company's plans challenged by the unprecedented halt that had been put on the freedom of movement of potential travellers.

It was not just the context of the Covid crisis that made it so difficult to clearly define whether the inauguration was a moment of hope or concern. Years before the pandemic, the blurriness between success and failure already defined my research. On the first day of the cable car's construction in summer 2018, Mr Frei oscillated between an unfettered optimism for the 'project of the century' and anguish for a future that constantly needed to be attended to, as he claimed that 'the biggest danger to tomorrow is today's success' (*Die grösste Gefahr von morgen ist der Erfolg von heute*). Complacency and inaction in the competitive tourism market could lead the valley to its demise, which would be a tragedy for the village and its inhabitants – all, more or less directly, tourism and JR dependents. If tourism had indeed 'made' the valley and region's history and success, Mr Frei and many tourism lobbyists (*Touristikers*) had spent the last years showing how quickly these achievements could be unmade. In Frei's discourses but also in the experiences of many locals, who had been increasingly pressured to take sides 'for' or 'against' the new infrastructure, promises of development repeatedly intersected with scenarios of decay and destruction.

The inauguration marked the first moment of certainty in a debate that had, until then, only embodied suppositions, speculations, approximations. The official opening of the cable car captured the precise moment when the cable car stopped being a future object and entered the realm of the present and past. In the pages that follow, I will examine what happened before the construction started, when the project's future was at stake, and the future of the whole valley was a topic of contention in a divided valley. As such, this chapter traces an archaeology of the future, revealing the ambiguous, almost schizophrenic, discourse of creation and destruction mobilized by Frei and his supporters, and its role in the making of the future of the valley. This approach in reverse, studying the past of a future-making, may seem more historical than futural. Yet, this chapter's focus derives from what characterizes the future in the presentist approach I adopt here: its omnipresence but epistemological inaccessibility for humans. The work, the energy mobilized by actors and/or institutions to make the future in an inherently uncertain context, are what remain graspable for the ethnographer (Nielsen 2011; Bear 2014a; Ringel 2018). The evocations of people who aim to make, orient, the future as an existential matter, in spite of its unknowability, all constitute the social practices geared towards the future that I wish to account for here.

All people are granted the power to temporally reason but also to mobilize temporal agency. They do so, for instance, to 'trick' the future, by 'individually and collectively attempt[ing] to modify, manage, bend, distort, speed up, slow down or structure the times they are living in' (Moroşanu and Ringel 2016: 17). Like most powerful capitalist actors, the JR company had worked to dominate futural discourse in order to foster its interests, revealing how institutions operating along the lines of capitalist modernity often successfully impose their futural projections. In the valley and elsewhere these projections have material consequences: the sprawl of infrastructure, but also the development of institutions and practices such as insurance, banking, loans, pension funds and financial speculation. By contextualizing JR's contested future-making and tracing its oscillation between alarming and hopeful messages, this chapter sidesteps the modern narratives of teleological development that naturalize growth as the only logical way forward. Instead, it reveals how the contestation of such ideals is constitutive of future-making – a reality in the valley but also beyond – in a global moment of increasing questioning of linear progress. Both the future and the present appear here as uncertain terrains that need to be conquered and navigated, even by powerful actors.

THE (ONLY) COMPANY THAT NAVIGATED CRISES

As Frei and his supporters claimed during the inauguration party 'ruined' by the Covid-19 pandemic, it was not the first time that the valley had faced unforeseeable events. The First World War had, as many in the valley described more or less historically accurately, abruptly put a halt to what was developing into a flourishing business, thereby forcing some locals to return to activities they had abandoned in favour of tourism, notably farming. Vestiges of infrastructures that did not recover from the war period were visible in the valley, maintained as heritage sites such as an elevator, built around 1900 – one of the first cable cars built in Europe.

But, contrary to what was implied during the inauguration, one did not have to wait until the pandemic to see another crisis emerge. Many locals in fact reported on an impressive succession of crises, some of which they only vaguely remembered. Since the two world wars, many 'abnormal' events had hit the valley before the pandemic: the oil crisis, the cold war, the Gulf wars, 9/11, the 2008 financial crisis, the outbreak of the avian flu virus, the Paris terror attacks, Brexit (with British investments suddenly retracted from the valley), and the increasing strength of the Swiss franc. In the opening talks of the tourism promoters and lobbyists in the many assemblies I attended, world events were regularly framed as difficulties, 'insecurities' (*Unsicherheiten*) that impacted, or could impact on, people's wanderlust (*Reisefreudigkeit*) in the Jungfrau Region. As one of the tourism lobbyists stated while pointing to the top of a Maslow pyramid[1] shown during a PowerPoint presentation, tourism was a luxury and not a 'need' for travellers, in contrast to the position of locals. This made the business and the village's position all the more fragile.

When I visited the valley post-pandemic in 2022 and met with hoteliers and locals I had last seen before the Corona outbreak, they unanimously described a feeling of tiredness and confusion, as they were no longer moving from one crisis to the next but found themselves at the intersection of multiple crises. One hotelier guiding me through his hotel shared the feeling of confusion with regards to new destabilizing global events impacting the valley, such as the lack of workforce following the massive lay-off of people working in the hospitality industry in 2020, as well as the situation linked to the war in Ukraine and the energy crisis this induced, all of which were combined with very high numbers of tourists visiting the valley after, but also during, the pandemic.[2] The exact reasons for things unfolding in such a manner were deeply puzzling and worrying to him because they were so hard to comprehend and disentangle. He described what has been increasingly coined as a condition of our times, and especially of tourism: one of 'polycrisis' where 'tourism capitalism continues to be imbricated in multiple and intersecting crises' (Bianchi and Milano 2024: 1; see also Henig and Knight 2023).

Crises are, of course, shattering events that draw one away from what is considered to be the normal, 'good', anticipated and linear life course or development of places and processes (Roitman 2013; Narotzky and Besnier 2014). But with the omnipresence of crisis-talk in the valley before and after the pandemic, as a place vitally but precariously linked to the rest of the world and its travellers' wanderlust, the resort escapes interpretations highlighting the abject character of crisis-events. Instead, the situation presented by my interlocutors drove me to consider crisis in the valley as a context, to borrow anthropologist Henrik Vigh's terms (Vigh 2008). Albeit speaking from a context of violence and trauma in Guinea-Bissau that is different from that of wealthy Switzerland, Vigh's work points to the importance of not considering crisis and uncertainty as isolated periods of abnormality or exceptionality. Instead of looking at crisis 'in context' and to posit the extraordinary nature of such moments, Vigh examines crisis '*as* context', marked by a sense of 'chronic' uncertainty (ibid.). Crisis thus has to be seen as 'a terrain of action and meaning rather than an aberration' (ibid.: 8), which, as I will show in this chapter, proves essential to the study of future-making in the valley.

In this volatile context, one crisis-object seemed to surpass all others in the region: the receding ski and winter tourism. This 'classic sector' of the Alpine tourism industry, as one *Touristiker* told me, which had made local history since the end of the nineteenth century (Barton 2008), was diminishing nearly everywhere in Switzerland and Europe (Schlegel, Caprarese and Roggo 2011; Müller-Jentsch 2017). In everyday narratives and in the media, winter tourism was described as the village's and the region's *Sorgenkind* or 'problem child' because of the unpredictable climate, the change in holiday patterns, and the consequences of financial crises weighing heavily on people's capacity to afford skiing holidays. As the face and driving force of JR, the oldest and most powerful transporting company and owner of the entirety of the skiing infrastructure in the valley, Mr Frei was particularly involved in such crisis-talk. When I tried to arrange to talk to him, as an unavoidable figure in the valley, JR's PR section suggested I should talk to Heinz instead – a JR executive in his fifties who would be able to provide information on both the region and the company. They told me that Mr Frei was 'very busy travelling through Asia' doing promotional work to acquire new markets. Meeting with him was a privilege that most in the valley did not have.

When I met with Heinz in a hotel restaurant he chose for the interview in spring 2018, he told me about his deep attachment to a village he had been born and raised in. For many of my interlocutors, Heinz was one of the few *Einheimische* who had worked at JR for his entire professional life and made it to the higher levels of the now very successful touristic company, increasingly influenced by 'outsiders' like Frei. For Heinz, like most villagers, tourism was essential for the village as it had transformed this 'very poor place made of beggars', as he knew from history books, into a wealthy touristic resort. But the transnational tourism economy was essentially shaped by difficulties. Heinz described the challenges of winter and especially ski tourism as a costly business, requiring massive investment in infrastructure and workforce around the ski lifts, for instance. The unpredictability of the weather, the decrease in the purchasing power of European 'classical' guests since the 2008 financial crisis, and the waning of interest in an expensive sport all made the winter season 'problematic'. Winter was now 'cross-financed' by the ever-extending summer season, during which predominantly 'new' types of guests from across the world came to visit the valley and the surrounding famous sightseeing platforms managed by JR.

Heinz, too, stressed the uncertain endeavour that tourism was, mainly with regards to the wanderlust of tourists – the village's main resource – by evoking an image: 'It is an immense puzzle of different things like social media, branding, Twitter, as well as service on the site TripAdvisor'. All these aspects, Heinz said, influenced individual tourism, something intrinsically difficult to control. In this challenging context, 'group tourists' from Asian countries, such as Japan, China, India, Korea and Taiwan, as opposed to individual tourists from Europe, had represented a significant advantage for their predictability, as they came, no matter the weather, to visit the valley's sightseeing platforms owned by JR. Since the Asian market had 'opened up' at the turn of the twenty-first century, JR had successfully attracted these new, predictable guests with lower prices. The company was involved in what Heinz described as a 'real price war' (*Preiskampf*) with the tour operators that organized such trips but also offered trips to similar destinations in Switzerland and Europe, all competing to welcome precious group tourists. However, this was a risky business too: Asian tourists were 'still people' and their 'crisis sensitivity' was high, Heinz added, confirming what many village dwellers were saying to me as they judged the reliance of the resort on such crowds as potentially damaging (see Chapter 3). The SARS outbreaks in the early 2000s had had particularly strong impacts on Japanese group tourists for instance. More recently, many Asian tourists were avoiding travel to Switzerland, including to the village, following various terror attacks in Europe and, for the Chinese especially, the Covid pandemic.[3]

In spite of these uncertainties, mirroring the insecure state of the world, Heinz retrospectively evaluated the history of his employer, almost surprised by its sudden successes. Whereas half a million visitors a year had been a major goal for the company for a long time, employees like him were astonished when the company reached it around the turn of the twenty-first century: 'We thought that's crazy back then and now it's over a million! It just doubled again in the last eighteen years, all this managed through group tourism'. State-of-the-art ski lifts, the renovation of sightseeing platforms, but also the whole winter season generally, had been financed by these group guests who had originally come to the valley between June and September and were now there almost all year round.

Building on these successes, JR increasingly became the most important player in the region. The company merged with and then tookover the village's municipal cableway infrastructure in the 2000s, which was in a critical financial state. While JR was previously active 'next to' the village, Heinz recalled how this repurchase gave a more active role to JR as the valley's economic core, incorporating locals who worked on the lifts in winter, often in combination with other activities such as farming in summer. Locals progressively had to work *with* JR, Heinz said, and 'everyone in the village was suddenly talking about which slopes were open and closed. . . . The relationship between the village and JR had become much stronger'. Contrasting with the narratives of unanimous success so often retold by *Touristikers,* who merged the interests of the JR company and the village, Heinz told me a more nuanced story: one of unequal benefit and pressure on the valley, generated by his company's rising success, mirroring the realities of many destinations around the world (e.g. Tucker 2005). Hoteliers, Heinz said, had trouble adapting to the accelerated pace of travel of the short-term guests JR pushed for, with the additional costs in personnel and resources they represented. The more comfortable times of long stays for regular guests (*Stammgäste*) were over, though longed for by many. The transporting company conquered the future and the uncertainty of the market, but it was not without its sacrifices for villagers.

At the end of our interview, Heinz once again referred to the long history of JR since the end of the nineteenth century. The company achieved the exceptional task of having 'survived three crises': the two world wars and the financial crisis of 2008. Such a history of success in spite of adverse situations, he claimed, revealed JR's capacity to make the best out of the worst conditions. Heinz was confident that the company was capable of overcoming the uncertain future: 'I think there are always possibilities to navigate the business through such crises'. Heinz further believed the new cable car project was 'crucial' (*matchenscheidend*). The new infrastructure would enable both the village and his company to realize profit, by raising attractiveness and promoting ski tourism to the disappearing 'classic' guests by bringing people to the slopes more quickly. Heinz professed his conviction that investing and building JR's new project was an essential way to make a future for his company and, to a certain extent, for the valley. To understand how such a belief emerged, we wind back to a summer evening of 2011, in the village's municipal hall.

THE NEW CABLE CAR

The sun is going down on the valley as the newly appointed CEO of JR, Mr Frei, dressed in a shiny grey suit and his darker hair pulled back, introduces his company's new project to the valley inhabitants.[4] The list of failed projects in the region is long, because of world wars, crises, or the locals' resistance. Yet, Frei has high hopes for his new cable car idea. In front of a sceptical audience after his 'surprise' project had been announced in the newspaper a few days earlier, Frei presents the new cable car as a means to secure not only his company's but the whole valley's future in an unstable context, particularly worrying with regards to the region's 'problem child': ski tourism. Thanks to the acceleration achieved by the new infrastructure, skiers will reach the slopes from the town of Interlaken in less than an hour; from the Swiss capital, Berne, within two hours; and from Zurich within three. Responding to widespread critiques that JR is mainly interested in attracting 'mass tourists', Frei argues that such a project, as well as the investments realized in recent years with new lifts and snow-making facilities, is proof that his company is prepared to invest millions in the winter business segment. Frei claims that everyone will profit from the project, but especially locals of the village – as tourism dependents who have been deeply involved with JR as the biggest touristic player and employer in the region.

Technically speaking, the new cable car is about new and faster connections. The project consists of a new terminal located on the valley floor with two new gondolas. On the western side of the valley, a bigger gondola will be built, next to the historical train connection that has remained unchanged since its construction in the 1910s. On the eastern side of the valley, JR will renovate and integrate an already existing cable car[5] into their project. When Frei delivers his speech in 2011, the historical train is the only way for individual tourists (like skiers) and group tourists to reach both the slopes and the sightseeing platforms located at an even higher altitude. These statements are not left unsubstantiated, as Frei enumerates figures and facts based on a study undertaken by a well-respected ski resort planning company. Listing the report's findings, Frei argues that the 'fill-time' (*Füllzeit*) – the time to 'fill' the resort with skiers – is too long because of the limited transport capacity of the current infrastructure. With the new project, the capacity could be increased to reach 3,600 visitors per hour.

The new gondola will solve not only this issue. Frei and his colleagues see the new project as the one and only chance to 'save' the existing cable car, as JR had done with the municipal cable cars and infrastructure a few years earlier. Built in the 1960s, in the booming years of skiing, this cable car had been one of the longest and most modern ski slopes at the time, but it now finds itself in a difficult situation. Its federal concession to operate is reaching an end in 2015, and to obtain a new authorization, the company has to invest and renovate on a massive scale. In a tensed financial situation since the outcome of the financial crisis of 2008, it is in an impasse. As a remedy to this situation, Frei proposes to integrate the existing cableway into JR's new project, which will reduce the investment needed from the company for the new concession. Facilitated by the new project, the CEO foresees a bright future of more guests and more 'quality' for JR, the company running the existing cable car, the village and the region as a whole, to be materialized in 2016 at a cost of 80 million Swiss francs.

By renovating and expanding the existing transport infrastructure, Frei predicts that the valley will 'enter the upper league'. The dissatisfaction of day trippers, especially skiers forced 'to wait before the beginning of the holiday experience', is particularly worrying because of its creation of 'disadvantages in the market'. This situation will lead to a predatory competition (*Verdrängungskampf*) among the existing ski destinations to which the village belongs, especially as ski tourism is 'no growth market', says the CEO. Alongside the existing cable car's problematic situation, Frei also addresses the hoteliers' concerns. Overnight stays are no longer a common occurrence in the region, leaving many hotel owners unable to renovate and maintain their infrastructures, which in turn decreases the earning value of their businesses. This 'vicious circle' is preoccupying for the entire destination, Frei claims. The new cable car is the only perspective of betterment left for all tourism dependents of the valley (i.e. everyone), as it will attract more tourists thanks to the rise in comfort represented by its unique technology. Ultimately, the project will enable the region to create new hotel beds, new job opportunities and new synergies. 'Something needs to be offered to guests . . . and time is running out', Frei adds, reminding everyone that the existing cable car's concession to operate will soon expire, and that other destinations are already providing better, faster and more comfortable travelling options for tourists. Through connection and acceleration, Frei, in the name of JR, presents a vision where local success can be achieved by behaving in accordance with the requirements of the global market.

ACCELERATIONIST PROMISES OF SECURITY

Almost ten years after Frei's introduction of the project of the century to the valley, I attend the inauguration of one leg of the project – the 'saved' cable car – in 2019. The smell of the fresh paint, the brand new cable cars and the proliferation of screens in what feels like an airport terminal impress me. Invited locals, too, pull out their cameras and phones to record a trace of this historic day. The next morning, I can read the ecstatic statements of the press enthusing about the '*krass modern*' quality of the new infrastructure.

As Brian Larkin put it, infrastructural projects like railways, roads, bridges, or even the world wide web are not merely material, they are also harnessed to particular affects, imaginaries, symbols or even 'poetics' (Larkin 2013,

2018). As promissory entities, infrastructural projects elicit and materialize modern ideals of betterment, progress and the common good that 'enchant' audiences (Harvey and Knox 2012; Abram and Weszkalnys 2013: 9–10). Promisors generally evoke an infrastructural future that 'exceeds the present' (Gupta 2018: 63), that will do more, better, to alleviate and develop the promisees' lives, relationships and conditions. For me on that day, like for many other villagers and commentators, enhanced speed and infrastructural growth indeed appeared as a reasonable, aesthetically pleasing solution to the resort's problems in a world of incessant movement. As soon as it happened, the inauguration was inscribed in the smooth history 'from farming to tourism' mobilized by so many.

It is precisely the taken-for-grantedness of infrastructural promises that makes them complex objects of analysis. As natural as they might seem, anthropologist Felix Ringel insists on the fact that all evocations of the future remain a mere representation of the inherently unknowable future that one invokes in the present (Ringel 2018). Promises of infrastructure follow 'scripts of modernity' (Gupta 2018: 62), mobilized by actors for particular purposes. The ski resort planning company's report presented the facts underpinning this script, encouraging an acceleration to reach a better 'fill-time' of the resort. But, as convincing as these seemingly objective data and rules of the market were, Mr Frei's future-making via speed participated in the speculative effort of 'channel[ing] the invisible and the unknowable . . . to realise incalculable possible futures' (Bear, Birla and Puri 2015: 388). The fact that such speculation rested on acceleration was no coincidence either. According to Koselleck's analysis of acceleration, it has been as a 'concept of historical hope' since the mid-eighteenth century, as opposed to the dreadful anticipation of the Last Judgement it had represented before (Koselleck 2004: 40–41).

But it was not simply such modernist obsession with speed that fuelled JR's accelerationist speculation. The promise of the infrastructure found an echo in the valley because it responded to a tangible experience of tourism's uncertainty and inequality acknowledged by all those involved in it. Heinz, as an employee of JR, but also Moritz, many hoteliers and other locals, were well aware of the declining numbers of skiers, the state of the existing cable car and the overcrowded slow train – all important features of the village that they unanimously believed had to be maintained and secured in a corrosive market. At the same time, they agreed that JR, the valley's 'milking cow' or 'lighthouse in the dark', had taken over nearly everything. The valley had become JR's property, instead of the opposite, as someone once told me cynically. Given the company's central role, or rather monopoly, over a region without alternatives to tourism, Mr Frei's promises had deeply affective undertones as the employer of the majority. Moreover, the maintenance of ski tourism was about more than just business; it was a symbol of a delicate symbiosis between farming and tourism activities. Ski tourism, with its lifts, stood for the vital and historical possibility of the 'genuine' local farmers being able to 'survive' in winter, one local woman said to me. It also stood for a Swiss sport tradition linked to the Alpine identity of the country and the region (Denning 2014; Engel 2021).

The CEO's promises also resonated among locals because of their own personal experiences of the infrastructure, as skiers themselves. Villagers, who were often born and raised in 'dynasties of skiers' (*Skifahrerdynastie*) as someone once put it, regularly opined that the project was necessary as they, as users themselves, knew what it meant to stand for hours in the crowded train to reach the slopes.[6] The slowness of the infrastructure made itself felt when locals visited ski resorts located outside the valley for work or leisure, in Zermatt or Austria for instance. 'You feel like crying (*Es kommen dir die Tränen*) when you are there, it is all so fast, smooth and easy, and much cheaper', explained Franziska, a local woman, following a trip to the Austrian 'rivals' where she painfully experienced the differences in infrastructures. The evocation of slowness and the responding promise of speed formulated by JR had powerful economic, but also affective and corporeal, resonance for many locals.

If the cable car was said to bring the region to the 'upper league' or the 'front line' of innovation, many village dwellers were seduced by the stabilizing potential of the new project. Innovation was perceived through the lens of continuity. JR's 'tradition of innovation', I was regularly told, had been initiated by JR's pioneering forefathers at the end of the nineteenth century, whose extraordinary visions had contributed to the present successes. 'Now we all benefit from their past investment and their visionary ideas', explained a local supporter and employee of the company I met as we were skiing and 'benefiting' from the infrastructure in the present. He urged locals to innovate in the same way as their ancestors had, to honour their forefathers' investments and to be inscribed in the same, continuous, even if accelerationist, teleology from farming to tourism. Contrary to many infrastructural projects that tend to 'evacuate the past' (Abram 2014: 104), the new cable car promised to break new ground and to stabilize the existing conditions, as products of a particular history.

For sociologist Rosa, this double bind is inherent to the functioning of capitalist modernity. Societies and institutions 'can only reproduce [their] structure through an *increase* of some sort – quite regularly, through (economic) growth, (technological) acceleration and/or higher rates of (cultural) innovation' (Rosa 2017: 440, italics in original). In the deeply competitive market of international tourism and in the historical resort of the Bernese

Highlands, such balancing acts between acceleration, growth or innovation on the one hand, and stability on the other, were inscribed in the valley's economic, social and material landscape and heritage.

The commitment of the 'project of the century' to securing local life, as the JR employee I met on the slopes said to me, was further guaranteed by the 'local' (*Einheimisch*) provenance of the money.[7] Contrary to what happened in the Swiss resort of Andermatt, where an Egyptian investor had injected money and thus gained power over everything, many supporters of the cable car believed JR's project to be particularly trustworthy because it had been financed through exclusively local capital. The supporters of the cable car I met told me they were also reassured by the presence of three locals on the JR board – another guarantee that the project would take the village's interests into account. At the same time, the fact that the innovative project was financed privately was a rare 'opportunity' that could not be dismissed. The JR company did so well that taxpayers' money was not needed, unlike other Swiss resorts, where the municipality was forced to invest in renovation of the transporting infrastructure in order to remain competitive. Researching reports and newspaper articles, however, I discovered that this misrepresented a more complex reality, with at least some (even if marginal) reliance on public investment.[8]

By interweaving the stabilizing and renewal potential of the project for the valley, JR's calls for acceleration appealed to supporters found in all segments of the local population: JR employees, communal and cantonal authorities, *Touristikers*, hoteliers, shop owners and various workers. From the beginning of the project, national political figures and heads of tourism lobbies had actively stood in favour of the cableway's potential for the future. While such a strategy of growth via speed may have been required for the valley to be able to reproduce its economic and social structure following Rosa (2017), it was also deeply contested.

PROMISE OR BETRAYAL?

When I arrived in the valley in 2017, the word 'new cable car' mainly evoked frustration among locals. For the previous six years, the topic of the future had profoundly divided the valley, as folk attempted to grapple with a situation described as 'standing still' (*Stillstand*) by Frei and newspapers but also villagers. Disconcerted supporters and opponents incriminated the other party for its actions against the common good, resulting in a stuck present. The valley's apparent immobility, however, did not imply passivity. Rather, it resulted in two opposing forces maintaining the village in an uncanny, seemingly tranquil but tensed atmosphere, enjoyable for tourists but bothersome for many locals, whose futures in the valley were rendered 'existentially problematic' yet unrealized (Ringel 2018).

No institution better epitomized the resistance against the 'project of the century' than the more than nine-centuries-old Alpine corporations or *Bergschaften* (singular *Bergschaft*), which Moritz, like many locals, described to me as features of the valley's uniqueness. Two of the seven local corporations whose task it was to look after Alpine pastures and common property resources[9] were impacted by the project, as the new route of the cable car was to cross their property. While the corporation located on the eastern side of the valley did not oppose the renovation of the existing cable car located on its territory, the members of the southern corporation expressed their stark refusal of the project from its very early stages, in 2011.[10] The opposition caused a stir in the valley and beyond, with regional and national newspapers largely reporting on what was seen as an affront to JR by local actors whose genuine belonging to the valley was uncontested. The farmers' corporations – the most ancient and traditional institutions in the village – were deeply anchored and respected in localist structures of power. Outside the valley's confines, international scholars had praised the role of these institutions, found in other Alpine valleys as well, in maintaining tradition and ecological balance through a long tradition of democratic[11] participation (Netting 1981; Tiefenbach 2013; Haller et al. 2019). Media headlines portrayed the mountain village as the theatre of a 'drama' unfolding between the traditional native farmers and the capitalist *Touristikers*, which presented parallels with the history of peasant uprisings that had formed the bedrock of the Swiss nation (de Capitani 2010).

Years later, when I met with village dwellers, especially supporters of JR, to talk about the opposition of the corporation members, many dismissed their views as outdated, narrow-minded and irrational. Yet, some active members at the time (mainly older native men) presented another side to the story. They had rejected the project out of fear for their subsistence with the new cable car project and its threat on the existing infrastructure, mainly the historical train. The corporation relied on the historical train's Right of Passage tax (*Überfahrtsrecht*) to continue the activities of the *Bergschaft*, such as working on the Alp to maintain the landscape to run its installations. At the same time, the flow of individual tourists brought by the historical train to the corporation-owned restaurants[12] was endangered with the advent of the new project. Olaf, an Alpine corporation member, explained that he feared that JR's new cable car would lead to the dismantling of the slow train in favour of the accelerated, more profitable cable car. Instead of securing local life, he saw the cableway as a form of rupture in local heritage. He and many of his colleagues feared that JR's need to accelerate, innovate and grow would be satisfied to the detriment of the corporation's land, its infrastructure, its business and its members' ways of life.

At the same time, by neglecting to talk more openly to the local corporations – as the *Touristikers*' traditional partners and well-respected symbols of the village, its traditions and its particularities – before submitting formal plans, Mr Frei seemed to have reactivated the historically inspired narratives of urban domination of the free mountain farmers. JR and Mr Frei, based closer to the lowlands, recalled figures of the 'foreign bailiffs' of the Middle Ages who long decided on the fate of the mountain people, as local Moritz had also implied (see Chapter 1). Olaf, the corporation member, confirmed in an irritated tone that 'We just read about it [the new cable car] in the newspaper'. Many corporation members felt that the future of JR was an imposition of the company's interested, globalized future over *their* long-standing traditions and farming history, *their* land, *their* activities and *their* efforts to preserve a unique environment and maintain its ecological and cultural equilibrium.

Alongside the local corporation's resistance, strict national environmental regulations protecting parts of the valley forced JR to rethink their plans. Yet, instead of downsizing the project, the company redoubled its efforts and designed a new, costlier project, finally budgeted at 510 million Swiss francs,[13] which circumvented the protected mountain area. Instead of dispelling the *Bergschaft*'s doubts, this move further raised tensions, and in 2014 the corporation officially voted against the realization of the new project.[14] Thanks to prescriptions inscribed in the corporation's very long history of preserving local resources, their opposition legally but also socially contested JR's legitimacy. Those who were already sceptical of the company's growing monopoly over the region found in this event another reason to oppose the project.

From the *Bergschaft*'s refusal onwards, my research partners remembered how lines of division between support and opposition increasingly materialized and 'cut through the village' (*das Dorf spalten*), as both Mr Frei and his opponents reported. In a climate of great tension, an official vote for the whole municipality took place, only a few weeks after the corporation's vote. Some opponents, from the corporations and beyond, described the impossibility of formulating critiques of the future proposed by powerful JR,[15] thereby questioning the sacrosanct value of democracy in the very heart of Alpine Switzerland. Comparisons between the valley and countries typifying totalitarianism and restrictions on freedom of opinion, such as Russia, Turkey and China, filled the newspapers at the time and were still part of the local discourses that opponents of the cableway mobilized when I was in the village.

Beyond the Alpine corporation's concerns and doubts, a whole faction of village dwellers wondered about the future impact of such a mega-project on their livelihoods, in 2014 and during the time of my field research. Next to those who opposed the gondola because their properties were about to be crossed by it, many locals who had witnessed the growing monopoly of JR also wondered whether the historical train would be abandoned for the sake of progress. Some (although a small number) shared their concerns over local jobs with me and wondered whether their positions would eventually become obsolete with the automatization and digitalization of tasks. Others uttered their doubts regarding JR's goal of 'saving' winter tourism, which Frei himself described as a 'no growth sector'. Perhaps, they wondered, the cableway was more about fostering a more profitable and more manageable 'mass tourism' for JR from which the village was so often said to 'not profit' (see Chapter 4).

In spite of these doubts, the 'vote of the century', as the media called it, took place in the village in 2014, a few months after the negative decision made by the corporation. Through a communal vote, the village's enfranchised population[16] was asked to put its own 'temporal agency' to use, to decide on how best to approach the problematic future of the valley. So many locals took part in the voting that the ice-skating rink was transformed into a giant municipal hall, with thousands of chairs for the voters, who were seated where tourists and the hockey club would normally skate. The vote was a success for JR with a 71 per cent acceptance rate from a participation rate of 56.7 per cent in the municipality, in spite of the corporation's opposition or the doubts cast by the project. This clear decision of the village community obliged the Alpine corporation to review its position, only a couple of weeks after their initial vote – another 'denial of democracy', as one of my interlocutors put it. The corporation finally accepted JR's cable car. But after the voting that many saw as an outrage,[17] nearly twenty private landowners, whose lands were to be crossed by the project, legally opposed it.

Beat was one of them. Going up one of the sides of the valley to reach his house, located more than forty minutes away from my home in the centre by foot and bus, I had already heard plenty of stories circulating in the media and within the village about his fight against the project. When I met this former JR employee and corporation member in his fifties, he recalled that he had read about the project of his then employer in the newspaper and learnt that the new cable car route was to cross his property. Beat's decision to legally oppose the project was motivated by personal reasons as the new project would disrupt his and his wife's tranquillity and retirement plans in a house he had inherited from his family and spent decades renovating. The new kitchen and the cosy living room in which we were seated were the result of his and his wife Hilda's efforts to transform hard-earned money, partly at JR, into their property, their *Heimat*. But his fight, he insisted, was also a fight *for* the village and against the new direction tourism development was taking under JR's leadership. The pressure at work, with an employer whose plans he challenged, reportedly became too great and Beat decided to resign. He recalled that other employees were disappointed when he left as they too were weary of JR's plans but did not want to express their views. Beat, a typical *Einheimisch* as he described himself, found it harder to stay silent when faced with what he saw as an upsetting situation and injustice. While I did not meet with many formal opponents of the project, the reasons behind Beat's appeal seemed at the heart of many opponents' narratives as they were reported in the village and the media. All highlighted a mix of personal concerns regarding their peace and properties, as well as more ideological justifications on a cable car that was to damage the valley's view and 'not profit' locals as dependents of tourism.

Behind the pledge for stability uttered by the *Touristikers*, prominently led by Mr Frei, Beat and villagers of various backgrounds wondered whether the future with the cable car did not in fact imply the valley's decline. Many recognized that JR's financially significant business was tied to the village's stability. However, they feared that the push for increasingly short-term tourist visits would attract visitors uninterested in supporting the very things that sustained the local community—such as small businesses selling regional products (see Chapter 3). Although traditionally committed to an economically liberal view – like in the rest of Switzerland, especially in the German-speaking part – many locals' concerns echoed David Harvey's critical description of the capitalist process of creative destruction. They worried that the pursuit of 'continuous innovation' was about to 'devalue, if not destroy, past investments and labour skills' shaping the valley's history and present (Harvey 1989: 105–6).

The destruction of the mythical Alpine landscape was at stake, too. Two national nature conservation organizations appealed against JR's project on the basis of its future effects on the environment, making use of particular provisions defined in the Federal Inventory of Landscapes and Natural Monuments of National Importance. Initally, Beat's hopes for a successful opposition to the cable car were motivated by his trust in these non-governmental environmental institutions – otherwise rather disliked in the valley because they were typical of a bureaucratic form of preservation of the Alps (Petite 2013). The NGOs argued that the cable car was to have a detrimental impact on an environment deemed of 'exceptional importance' in the national inventory. The new technology, the environmentalists stated, would disrupt the local fauna and flora of the Alpine moors because it involved building giant pillars in the middle of previously 'untouched' (but not protected) areas of the region. But the NGOs mostly insisted, via media coverage, on the aesthetic impact of such an infrastructure on the landscape and the *Dreigestirn* mountains as a 'unique' panorama in Europe.[18]

The local, regional and even national newspapers became the arena of intense discussions between environmentalists and JR, mostly represented by Mr Frei, regarding the cableway's route and the height of the new pillars needed in order to maintain the integrity of the view. For the NGOs, the 'visual impact' (*visuelle Beeinträchtigung*) on the mountain landscape implied a more symbolic critique against JR, whose employees were committed to realizing a future made of 'mass' instead of 'class' tourism, partly echoing the concerns of the Alpine corporations and many locals. The transport company and its supporters retorted that the integrity of the view was maintained, a decision confirmed in 2017 by the Federal Commission for the Protection of Nature and Cultural Heritage.

Far from being sealed, the fate of the cable car remained uncertain long after its acceptance by a majority of the population in 2014. By mobilizing their temporal agency thanks to legal frameworks, Alpine corporations, landowners and NGOs tried to stop the realization of what many deemed an undesirable future. The repercussions of these contestations were predicted by the media and many *Touristikers* to provoke 'decades of standstill'. It was only halfway into my first period of fieldwork, at the beginning of 2018, that Mr Frei was able to discard the majority of the private appeals against the project. The last agreements with the environmental NGOs and the authorizations to build were finalized in summer 2018 following an agreed reduction in the size of the pillars, along with changes in the design of the cabins and a diverse part of the infrastructure zones. The company also reiterated its intention to maintain the historical train for the next fifty years.

The construction process could finally start after more than four years of delay and was inaugurated in December 2020, effectively reducing the travelling time of tourists by half. If touristification was said to have pulled locals out of farming and poverty in the past according to a widely shared belief in the teleology of modernity, my research was shaped by a strong contention of the future. Such stories of conflict and uncertainty, which seem to have accompanied many infrastructural projects in the valley,[19] have become crucial to retell as they radically question our retrospective understanding of technological progress and the common good.

A CHANGE IN TONE

When I arrived in the village in 2017, JR had been showing record numbers of visitors for five consecutive years and the existing cable car's concession to operate had been extended thanks to its future integration into the new project. Yet, cantonal authorities[20] had refused a construction permit based on technicalities. The company was both thriving and challenged in its authority by the appeals and this refusal. Many village dwellers admitted being tired of the situation, along with the tensions and divisions that the project had caused before its construction had even started. In the media and assemblies that I attended, Mr Frei seemed to embody this mix between success and setback. He admitted feeling impatient, worried or irritated one minute, and overtly optimistic and confident the next, as he prophetically claimed once again that the project would make the village enter the 'upper league' – the same state I saw him inhabit three years later as I watched the inauguration from my computer during the Corona crisis.

The tensions that existed before the appeals' dismissal were indeed palpable as Mr Frei issued a warning to the valley dwellers with the motto, 'It's later than the eleventh hour' (*Es ist fünf nach zwölf*). The sentence 'Standing still is regressing' (*Stillstand ist Rückschritt*) had become part of his everyday vocabulary, stressing how time passing without the new cable car was detrimental for everyone, even if the fragility of the situation remained invisible, 'for now'. If capitalism is characterized by its 'mixed utopic and dystopic representations of time' (Bear 2014b: 78), Frei clearly evoked the second as he told various audiences in person or via the press that the company was losing money because of the appeals. The region could soon lose its leading position in the global tourism market, where rivals continued to innovate and improve upon their increasingly fast, more comfortable tourist infrastructure. For those who were to read JR's current success as a sign that the cable car was not needed, Frei argued for the need to perpetually invest in the future because 'tomorrow's biggest threat is today's success'. Other ski resorts and sightseeing spots were already catching up. On the day of the much-awaited construction start, Frei attended with a giant clock under his arm and illustrated his speech by moving the hands of the clock: with the construction start, the company and the region finally caught up with time and compensated for their delay so as to be 'on time'.

Such crisis-talk was already present in 2011 as the background against which the resort needed to develop in order to remain one of the best if those who depended on it to make a living wanted to stay in the race. But the pervasiveness of uncertainty in the tourism business – a fact that JR had always brilliantly navigated, according to Heinz – now came to the fore with strong negative undertones. The future of the valley was no longer achievable by optimistically following the tradition of innovation inherited from the forefathers of tourism. It now had to be *dominated* according to the predatory rules set by a market where others elsewhere were already better in responding to the unpredictable tourists' wanderlust and their quickly changing desires. This situation could turn into a disaster if nothing was done, Frei warned. On the day of the project's opening in 2020, Frei himself retrospectively explained in the media that 2016 marked a particularly difficult moment, with an increase in appeals against the cableway. One moment in particular marked this switch to a resolutely darker narrative, with the publication of a video entitled 'Jungfrau Region 2026' on the YouTube channel of the company.

JUNGFRAU REGION 2026

The video, set in 2026, opens with a boy calling out, in local Swiss-German dialect, 'Grandpa, Grandpa!' (*Grosätti, Grosätti*) as he runs across the screen.[21] The boy joins his grandfather, who is sitting on what seems to be a bench in the middle of an Alpine field surrounded by famous mountain peaks. The bench is in fact a dismantled chairlift seat resting on the ground, detached from the rest of the ski lift to which it would be normally harnessed. The child looks confused and asks what this 'funny seat' is. 'Ten years ago, long before you were born, we could still ski here', responds the old man. Pictures of winter landscapes with crowded restaurants, slopes and ski lifts appear in a flashback as a sad song fades in. Displaying the promotional material that could be seen on billboards and screens, and at meetings and assemblies during my fieldwork, these images attest to another time in the video: 'Tourists came to us from all over the world, . . . everyone had a job, we were doing well . . . no wonder with these beautiful mountains!', says the old man, talking over images of rosy sunsets above the snow-capped landscapes. 'We thought we had everything. We became lazy and we dismissed every new idea. Why change something like that?' asks the grandfather.

The man's tone of voice changes as the idyllic destination's images now change to black and white, and slow motion: 'We thought it was enough to continue as we had done. [But] things went bad'. The screen is now filled with a map of the resort, with the connection lines branching out to indicate what ski lifts, cableways and railways serve the valley. As the old man speaks, the lines quickly fade away to leave a blank, empty space made of an almost untouched Alpine valley, with only the name of the region's villages and a few roads. Talking over images of the empty, shut-down tourism office and ski lift installations turning into ruins, the man adds: 'Nothing stayed as it was before. Everything went up the creek (*alles ist bachabgange*). I could keep the chairlift seat, and now it reminds me how stupid I was'. Stroking the boy's head, the grandfather concludes sadly, 'If only I had not hindered the new cable car ten years ago', contemplating what we now know is a valley in ruins. The following statement appears: 'It would be a pity if the new cable car project were to be stopped, because it secures the successful touristic mid- and long-term future of the whole region as a top, all-year-round destination in Swiss tourism. It strengthens the competitive edge of the [*Dreigestirn*] as a globally known lighthouse (*Leuchtturm*), and helps the winter sport destination reach a leading position in the international business competition. More value creation and safe jobs – we need the new cable car!'[22]

As if shot in the future, the video made the present into the past and combined different temporal frames within only two minutes. In what Anna Tsing (Tsing 2004) describes as the capitalist 'economy of appearances', which relies on the display of 'economic' but also 'dramatic performances' or 'spectacles' to attract capital and investment (ibid.: 118), I was surprised by the fact that one of Switzerland's most successful touristic companies purportedly evoked its own ruination within a decade. While JR was thriving and had regularly put forward the solidity of their success in promotional material, this video seemed to make a rather counterproductive claim. It showcased the uncertainty and the fragility of Mr Frei's own endeavours, and drew on an uncanny proximity between failure and success.

Upon its release, the video was received very differently in the village and beyond: while some – mostly *Touristikers* and supporters – were convinced by its urgent message, others judged it to be manipulative, simplistic and overly emotional. I myself was puzzled by the video but nonetheless recognized some of the tropes evoked by my interlocutors. The video spoke about the economic but also the affective attachment of village dwellers to tourism, its essential role in making them 'something' as opposed to them being 'nothing' without it. It also resonated with Heinz's account of the great power of JR over the valley: without the company's project, one could expect nothing less than the ruination of the whole valley. It further revealed the consequences of an uncertainty that Heinz explained to be intrinsic to the business of tourism, relying on constant innovation to avoid rapid and dramatic failure. To decipher the meaning of the video and its reception among valley interlocutors, I needed to dig deeper, to uncover its temporal structure.

A FUTURE ANTERIOR IN RUINS

In the video, the grandfather appears as the all-knowing judge of the present of the village. He looks back to the past – the viewer's present – with nostalgia as he reveals his regrets. The old man admits having been naive, selfish and responsible for his own ruination and the collapse of the whole region. He has caused a rupture in heritage and history highlighted by his grandson's lack of knowledge of what used to shape the valley's economy, i.e. tourism. The affective charge of this structure is strong: an old man harming his kin and community, and himself admitting his wrongdoing but unable to repair what is now forever lost. The video also allows its viewers to uncannily reflect on and evaluate a present in which they are embedded, from the position of someone located in the future.

In her work on the sociopolitical use of grammatical tense, Elizabeth Povinelli coins a term for the video's temporal structure – the 'future anterior': a 'future interpreter . . . is seeing what is in front of our eyes but outside our field of vision' (Povinelli 2011: 3). She demonstrates that future anterior is an important temporal structure of late liberalism because it allows for the justification of sacrifices in the present to move people to act according to a future point in time, be it the Apocalypse, posterity or history, broadly speaking. In other words, the future anterior allows humans to powerfully overcome the uncertainty of the future to act in the now (see also Caduff 2014).

The future anterior of 2026 tells us that the most convinced opponents of 2016 will soon regret their narrow-mindedness, stubbornness and stupidity. They will be confronted with the tragic consequences of their actions on the valley and their loved ones, who will, inevitably, judge them. The absence of the new infrastructure will equate with the (very) rapid destruction of everything that makes the present: tourism, public and private transport infrastructure, success, crowds. Everything will go 'up the creek' (*bachabgange*) because of opponents who appear capable of influencing the flow of time, 'tricking it' (Moroşanu and Ringel 2016) for their own selfish but irrational interests, at odds with JR's goal of making for a common future of stability for the valley's population. By revealing the dangers posed by inaction in an uncertain, competitive present, the video makes the invisible visible, in a moment where the company was financially successful but contested in its plans.

Whereas I initially perceived the tone of the video as overtly dramatized, many locals did not discard it as quickly. One could disagree with the artistic direction of the video, but its message was essentially deemed realistic because it contained some undeniable truths about the valley's context and the lives of its inhabitants. The overbearing importance of tourism, the monopoly of JR in the region, and a sense of pervasive uncertainty all shaped their slippery situation as mountain dwellers living off the 'business of foreigners'.

Anna, an *Einheimische* in her early twenties who had been employed at JR since her sixteenth birthday, told me she believed the scenario of ruination without the cable car to be credible. This perception was rooted in her experience of a company that not only dominated the valley but also had a central place in her life and that of her relatives. Like the majority of her friends, her father, her brothers and her grandfather had all been employed by the valley's 'milking cow'. As a young local native, Anna said confidently that she knew exactly what she was dependent on, contrary to the naive opponents. Dressed in expensive skiing clothing after an afternoon on the slopes on her day off, she explained that, in her opinion, old men who had 'never left their village' resisted everything new. If the latter stopped JR – the valley's only actor capable of safely bringing the valley into the future in deeply uncertain times – in its attempt to innovate, she feared the potential for the complete disappearance of tourism, its infrastructure and history, and the impossibility of a viable life for mountain dwellers. For Anna, not doing 'anything' and 'continuing like for the past hundred years' in a world of acceleration, as opponents wanted, was narrow-minded and backward. Echoing the sentiments of the video, she believed such an attitude would lead the village to go 'back' to isolation and poverty. This image evoked a return to the farming (*bure*) activities her ancestors had performed, and the spectre of outmigration. Without the cable car and, eventually, tourism in general, she imagined how her generation would have to leave the village to seek better futures in the urbanized lowlands – a place she disliked but where jobs would remain available, contrary to the non-touristic mountainous areas. Going back to pre-tourism times evoked a past when poor locals left the valley for North America in the eighteenth century (see Lerch 1909).

Not everything that belonged to the farming past was grim. Anna believed it was important to stay connected to her roots, and to honour her ancestors by maintaining the typically Swiss traditions such as yodelling. In our interviews and meetings, she, like Moritz, historicized this past to defend traditional Swiss and local values in an anticosmopolitan, antiprogressive agenda against newcomers to Switzerland and the village *in the present*. Yet, coming back to these farming roots and the poverty was not a possibility for Anna, who admitted it would destroy her sense of how to be in the world and live a dignified life. She wanted comfort, leisure time and travel, along with the ability to move around like any citizen of the twenty-first century. Supporting JR's project did not just mean supporting her employer's plans. It meant belonging to a certain type of *Einheimische* that was for innovation and the conditions that would guarantee her a future worth living in the mountains. The old, stubborn mountain dwellers who opposed the project were the 'enemies of tourism', the 'primitive' autochthons, to take up on Geschiere's definition of autochthony's ambiguity (Geschiere 2009) discussed in the previous chapter. I heard Anna's reproach being reiterated countless times in the valley by villagers of different backgrounds.

Beat was regularly the target of such accusations. Villagers like Anna implied that he was either backward or motivated by 'greed' and the financial arrangements he could obtain from JR rather than an actual 'ideological' problem with the new project. Anna's family, with whom I once shared lunch, described Beat as 'stubborn', 'responsible' for the delays in the project, for the money JR lost and the uncertainty he provoked, but also for his exclusion from the village life, which he had 'built for himself'. Beat was a 'typical *Ureinheimische*', but not the kind I should talk to. Instead, Anna's father claimed he should be completely 'expelled from the community'. If both Anna's family and Beat were all natives, Beat was the abject, 'ignoble' version of the autochthon, while Anna's family members explained that they came from the 'sunny side' of the valley, and had a more enlightened disposition and prestigious status (Berthoud 2001: 85; Geschiere 2009: 11).

When I met Beat for the first time in 2018, he had just signed an agreement with JR precisely because of the humiliation he had been subjected to over the last years. He and his wife, Hilda, had asked me 'not to talk about the cable car too much', as they both were exhausted by 'all the drama', as they called it. Under the company's and the locals' pressure, they could no longer fight, they said to me. JR had regularly evoked the possibility to expropriate them, which the company would legally be able to do under provisions prescribed by the Federal Law on Cableways and Transport (*Seilbahngesetz*, SebG) and the Federal Law on Expropriation (*Bundesgesetz über die Enteignung*, BiG), defining cableways as contributing to the 'common good' (BiG Art.1). The CEO insisted in the media that this was the 'worst case' (*schlimmster Fall*) scenario as it would take time, but it was possible. During my first visit to their home, which lasted three hours, as well as in subsequent visits, Beat and Hilda spoke almost exclusively about the cable car, revealing just how much the project dominated their lives.

The video 'Jungfrau Region 2026' found a particular echo with Beat, who clearly showed resemblance with the main protagonist, as an older, *Einheimische* opponent. Expecting him to be angry at such a depiction, I was surprised when he said he was most hurt by those involved in making the video. One of them in particular used to be a friend, and had even taught Beat how to ski as a kid. This fellow *Einheimische* used to be someone that he admired, but now he had betrayed him, 'like so many others', by participating in a campaign that had depicted people like him as selfish and backward. 'It really hurts', Beat said quietly. The video epitomized not only an attack on his person but also what he saw as JR's manipulation of the locals' complete dependency on tourism and the company. This had the consequence that fellow locals came knocking on Beat and Hilda's door, telling them that if the existing cableway stopped it would be because of their opposition. 'If it stops, we are guilty', Hilda said in a sad voice before telling me how unjust this situation was. 'I am not a mechanic or an engineer; it's not my responsibility if it stops, but for them [the locals], it is', she shrugged.

In attempting to save his house, his *Heimat*, Beat explained that the project had 'ruined his life' from many aspects: the project was going to be built, the fight had cost a fortune, his pre-existing medical conditions had worsened with the stress, and he felt extremely disappointed. When he later donated the compensation money he had received from JR, following his agreement with the company, to local clubs, he was not even thanked. As one of these club members told me, it was too late to show generosity after he had acted so selfishly by opposing the cable car in the first place.

Beat's demonization in the village resulted less from his 'backward' character than from his role in imposing what Anna and many others saw as a forced *return* to the past for all village dwellers. The past was desirable when over, but abject when it signified the imposed, imminent loss of the present, and the decay of everything one had known and loved. The ruined future of a successful infrastructure featured in the video implied a painful state, a collapsed wealth and a dramatic loss of stability. In Gupta's words, these ruins and those who provoked them were made into extraordinary, abnormal, worrying objects that contradicted the 'script of modernity' (Gupta 2018: 62).

Beyond the strong symbolic power of ruins, I believe that the scenario of demise felt credible because it also recalled people's material proximity with ruins in the present. Abandoned, empty hotels and the deteriorating state of the infrastructure (such as the existing cable car) stood for the uncertainty of the tourism industry, as well as the dramatic consequences of people's incapacity to adapt to the latter. They were the result of people having 'slept on their laurels', of having mismanaged, or failed in one way or another, for reasons most of my interlocutors could not fully grasp. Such a present revealed the increasing pressure they felt, to be doing more to attract tourists, as they were no longer able to do business 'as before'.

Visiting meetings held by local associations, I could see how locals, especially older ones, wishing to host tourists in their holiday rentals felt the pressure to respond faster, set up digital guest cars, learn English, learn about their rivals in the other valleys. This pressure was embodied by tourism officials who, on one occasion especially, forcefully told them that there was 'no turning back' in tourism development, and that they had to comply with the accelerating and changing rules of the market. The consequences if they did not do so did not need to be spelt out; they were in the video, in Mr. Frei's speeches, and across the street from where the meeting was held, where a ruined hotel stood as a painful reminder of someone's maladaptation to tourism's demands.

The video then needs to be interpreted within a context shaped by concerns similar to those described by anthropologist Joseph Masco in his work on the 'business of survival' and the affective economy of nuclear destruction in Cold War United States (Masco 2008). Based on fictional evocations of the future through reconstitutions or films, he shows how a future of atomic annihilation and ruination was displayed to govern citizens. Nuclear fear, Masco states, was 'inoculated', making 'every citizen into a potential victim' (Masco

2008: 268), to advance particular political agendas in the present. Because of its evocative and tangible power, I suggest that the images of ruination troubled and spoke to valley dwellers, who saw themselves as potential victims of a future of dispossession without JR and tourism, as dependents of both.

CONCLUSION

Following three years of recession in the aftermath of the Corona crisis, the cable car has now enabled JR – and the valley, according to many – to 'recover' in record-breaking time. Innovation, the latest figures show, was the 'right' decision. In this context, as fervent JR supporters I met before and after the construction implied, these successes were the sign of the naivety of opponents, who were proven wrong and forced to admit their misjudgement. I have no doubt that the cable car will soon be incorporated into the smooth teleology 'from farming to tourism' (see Chapter 1). Resistance, framed as backwardness and stupidity, however rarely makes it into history books following a modernist narrative. Yet, as a local once told me, the 'wounds of the divisive project, not just on the landscape but in the village, will take time to heal'. These wounds are likely to be reactivated when confronted with the next project – one that is already under way in 2025, at the time of redaction of this book, with the renewal and extension of yet another important part of the village's transport infrastructure.

But what the controversy of the new cable car also shows is how contested modern and capitalist future-making remains, even as it seems to propose solutions that 'enchant' audiences. To make the future according to its interest and to impose it as the one and only desirable one for valley dwellers, the company JR was forced to negotiate and respond to opponents' fears and concerns. Decisive in this process was not just JR's resorting to modern promises of security. It was also – and increasingly so – done via darker evocations of ruination and the total destruction of tourism in the valley, i.e. *everything*.

If it is widely known that instability participates in the (re-)making of capitalist structures (Harvey 2007), what is surprising here is that the company itself chose to reveal the uncertainty of the market and their own activities – things they normally claimed to dominate. Crisis as a 'context' (Vigh 2008) – manifest in the unpredictable tourists' wanderlust, the global situation, the receding of ski tourism, a history of past poverty and the lack of alternative means of subsistence in the mountains – was used to appeal to the complete dependency of locals on tourism and on the powerful company to push people to support its project. Such a terrain of uncertainty and its associated meanings was used to formulate appealing promises and evoke threatening scenarios. If, as Heinz said to me, JR made its successes 'in spite' of a difficult context, my findings show how the company could make the future 'thanks' to this context and its evocation for villagers who depended on tourism. Rather than opposites, promise and threat, infrastructure and ruination, were two sides of the same capitalist future-making.

NOTES

1. The Maslow pyramid is a visualization of the hierarchy of needs. The base of the pyramid represents the most fundamental, basic and physiological needs, while the upper part represents areas of self-actualization that can only be satisfied once the most basic needs have been met.
2. Although the overall number of tourist visits significantly decreased in 2020 due to various measures, many local owners of hotels and restaurants explained that they were surprised by the very high number of guests, mostly from Switzerland, during the pandemic.
3. https://www.swissinfo.ch/eng/business/chinese-tourist-flows-to-switzerland-will-not-fully-recover-before-2026/48863960 (last accessed 14 April 2025).
4. This section is based on local newspaper articles and reports between 2010 and 2018, as well as participant observation in various assemblies during my fieldwork between 2017 and 2018. To grant the anonymity of village actors and my interlocutors, I have refrained from citing my sources here.
5. Of which JR held 30 per cent of the shares already.
6. The ski club was an important part of the village life, and skiing competitions part of the school curriculum, as has been the case until recently across Switzerland. See Engel 2021.
7. Many JR supporters also claimed that a great number of villagers had shares in the JR company, which was another reason to accept the cable car. I was not able to find figures on this.
8. Out of 510 million Swiss francs that the project finally cost – a significant increase from the original estimate – at least 320 million came from JR's private investments. 150 million came from lease and state investment. The company of the existing cable car spent 35 million, mainly financed by a leasing agreement and a loan with a private bank; and 100 million was invested by a public railway company that had been closely collaborating with JR for the past century. 20 million came from the state and canton of Bern as a means to develop the touristic projects of the highlands. These figures come from reports and media articles. I refrain from citing sources in order to align with my anonymization strategy.
9. Such as 'grazing land . . . forest, roads, water bodies' too (Moser and Straub 2016: 3).
10. I heard of another earlier project planning for a new construction that had also been previously rejected by the eastern corporation. The 'simple' renovation of the existing infrastructure – although implying the replacement of the new line by bigger gondolas – was accepted.
11. This tradition of democratic participation, which resonates with the Swiss state-building enterprise, needs to be situated. Democratic rights within the corporations were in fact only granted to men in the valley until the 2000s, based on

property and inaccessible to foreigners, like in many other corporations (Haller et al. 2019).

12. Located next to the railway tracks.
13. Instead of the initial 80 million.
14. According to the corporation's regulation, two-thirds of the votes were required to have the project accepted.
15. Doctors, hotel owners and private landowners of the village who were against the project expressed their concerns as they reported on the increasing pressure on their villagers' shoulders and feared for the democratic climate in the village. In a petition, more than 900 of the 2,600 inhabitants eligible to vote insisted on the importance of conducting a postal ballot instead of the regular communal assembly in order for them to exert their right to democracy without being subjected to outside influence. This was judged 'unnecessary' by the regional prefect, and even went against the municipal regulations stipulating the need to hold an assembly regarding infrastructural projects.
16. Like in most of Switzerland, the conditions to vote were: being eighteen, having one's residency in the village, and being Swiss. I will come back to this in Chapter 5.
17. The democratic values were further questioned by the fact that the assembly of voters was promised new things a few minutes before the vote. In addition, the fact that the corporations were made to re-vote so shortly after their refusal was deemed undemocratic.
18. I here again refrain from citing my sources (newspaper articles dating of 2017) to protect the anonymity of my research informants.
19. The historical train built by JR in the 1910s was also deeply contested by the local population, as historical accounts show (Moser 1997).
20. Notably, the Directorate of Justice, Municipalities and Churches (Justiz-, Gemeinde- und Kirchendirektion, JGK).
21. https://www.youtube.com/watch?v=oBk0-Q4ZU30 (last accessed 15 April 2025).
22. My own translation from German.

REFERENCES

Abram, Simone. 2014. 'The Time It Takes: Temporalities of Planning', *Journal of the Royal Anthropological Institute* 20(S1): 129–47.

Abram, Simone, and Gisa Weszkalnys. 2013. *Elusive Promises: Planning in the Contemporary World*, Vol. 11. New York: Berghahn Books.

Barton, Susan. 2008. *Healthy Living in the Alps: The Origins of Winter Tourism in Switzerland, 1860–1914*. Manchester: Manchester University Press.

Bear, Laura. 2014a. 'Doubt, Conflict, Mediation: The Anthropology of Modern Time', *Journal of the Royal Anthropological Institute* 20(S1): 3–30.

———. 2014b. 'For Labour: Ajeet's Accident and the Ethics of Technological Fixes in Time', *Journal of the Royal Anthropological Institute* 20(S1): 71–88.

Bear, Laura, Ritu Birla and Stine Simonsen Puri. 2015. 'Speculation: Futures and Capitalism in India', *Comparative Studies of South Asia, Africa and the Middle East* 35(3): 387–91.

Berthoud, Gérald. 2001. 'The "Spirit of the Alps" and the Making of Political and Economic Modernity in Switzerland', *Social Anthropology* 9(1): 81–94.

Bianchi, Raoul V., and Claudio Milano. 2024. 'Polycrisis and the Metamorphosis of Tourism Capitalism', *Annals of Tourism Research* 104: 103731. Doi: https://doi.org/10.1016/j.annals.2024.103731.

Caduff, Carlo. 2014. 'Pandemic Prophecy, or How to Have Faith in Reason', *Current Anthropology* 55(3): 296–315.

de Capitani, François. 2010. 'Peuple des bergers', *Dictionaire Historique De La Suisse*.

Denning, Andrew. 2014. *Skiing into Modernity: A Cultural and Environmental History*, Vol. 3. Oakland: University of California Press.

Engel, Simon. 2021. 'The Whole Country Is Skiing! The Whole County. . .?' *Nationalmuseum.Ch*, 13 January 2021. Retrieved 22 April 2025 from https://blog.nationalmuseum.ch/en/author/simon-engel/.

Geschiere, Peter. 2009. *The Perils of Belonging: Autochthony, Citizenship and Exclusion in Africa and Europe*. Chicago: University of Chicago Press.

Gupta, Akhil. 2018. 'The Future in Ruins: Thoughts on the Temporality of Infrastructure', in Nikhil Anand, Akhil Gupta and Hannah Appel (eds), *The Promise of Infrastructure*. Durham, NC: Duke University Press, pp. 62–79.

Haller, Tobias, et al. 2019. *The Commons in a Glocal World: Global Connections and Local Responses*. London: Routledge.

Harvey, David. 1989. *The Condition of Postmodernity: An Enquiry into the Origins of Cultural Change*. Oxford: Blackwell.

———. 2007. 'Neoliberalism as Creative Destruction', *Annals of the American Academy of Political and Social Science* 610(1): 21–44.

Harvey, Penny, and Hannah Knox. 2012. 'The Enchantments of Infrastructure', *Mobilities* 7(4): 521–36.

Henig, David, and Daniel M. Knight. 2023. 'Polycrisis: Prompts for an Emerging Worldview', *Anthropology Today* 39(2): 3–6. Doi: https://doi.org/10.1111/1467-8322.12793.

Koselleck, Reinhart. 2004. *Futures Past: On the Semantics of Historical Time*. New York: Columbia University Press.

Larkin, Brian. 2013. 'The Politics and Poetics of Infrastructure', *Annual Review of Anthropology* 42(1): 327–43.

———. 2018. 'Promising Forms: The Political Aesthetics of Infrastructure', in Nikhil Anand, Akhil Gupta and Hannah Appel (eds), *The Promise of Infrastructure*. Durham, NC: Duke University Press, pp. 175–222.

Lerch, Ernst. 1909. 'Die Bernische Auswanderung Nach Amerika Im 18. Jahrhundert', *Blätter Für Bernische Geschichte, Kunst Und Altertumskunde* 5(4).

Masco, Joseph. 2008. '"Survival Is Your Business": Engineering Ruins and Affect in Nuclear America', *Cultural Anthropology* 23(2): 361–98.

Moroşanu, Roxana, and Felix Ringel. 2016. 'Time-Tricking: A General Introduction', *The Cambridge Journal of Anthropology* 34(1): 17–21.

Moser, Patrick. 1997. *'So wird die Jungfrau zur Demoiselle gemacht': Projektierung und Bau der Jungfraubahn*. Zurich: Chronos.

Moser, Ruth, and Ueli Straub. 2016. *Swiss Land Governance*. Lindau, Ch: Agridea.

Müller-Jentsch, Daniel. 2017. 'La crise du tourisme de montagne en Suisse'. Blog. *Avenir Suisse*.

Narotzky, Susana, and Niko Besnier. 2014. 'Crisis, Value and Hope: Rethinking the Economy. An Introduction to Supplement 9', *Current Anthropology* 55(S9): S4–S16. Doi: 10.1086/676327.

Netting, Robert M. 1981. *Balancing on an Alp: Ecological Change and Continuity in a Swiss Mountain Community*. Cambridge: Cambridge University Press.

Nielsen, Morten. 2011. 'Futures Within: Reversible Time and House-Building in Maputo, Mozambique', *Anthropological Theory* 11(4): 397–423.

Petite, Mathieu. 2013. 'Les montagnards face aux ecolos: une énième déclinaison au travers de l'initiative Weber', *Journal of Alpine Research* (Hors-Série).

Povinelli, Elizabeth A. 2011. *Economies of Abandonment: Social Belonging and Endurance in Late Liberalism*. Durham, NC: Duke University Press.

Ringel, Felix. 2018. *Back to the Postindustrial Future: An Ethnography of Germany's Fastest-Shrinking City*, Vol. 33. New York: Berghahn Books.

Roitman, Janet. 2013. *Anti-Crisis*. Durham, NC: Duke University Press.

Rosa, Hartmut. 2017. 'Dynamic Stabilization, the Triple A. Approach to the Good Life and the Resonance Conception', *Questions De Communication* 1: 437–56.

Schlegel, Nicole Brändle, Marco Caprarese and Stefan Roggo. 2011. 'Le tourisme en Suisse – Stations de ski en concurrence', in Martin Neff (ed.), *Swiss Issues Branches*. Zurich: Credit Suisse.

Tiefenbach, Marianne. 2013. *Alpkorporationen-Traditionelle institutionen nachhaltiger Landschaftsentwicklung: Das Beispiel der Bergschaften Grindelwalds im Kontext aktueller gesellschaftlicher Herausforderungen*. Grindelwald: Sutter Druck.

Tsing, Anna Lowenhaupt. 2004. 'Inside the Economy of Appearances', *Public Culture* 12(1): 115–44.

Tucker, Hazel. 2005. *Living with Tourism: Negotiating Identities in a Turkish Village*. London: Routledge.

Vigh, Henrik. 2008. 'Crisis and Chronicity: Anthropological Perspectives on Continuous Conflict and Decline', *Ethnos* 73(1): 5–24.

CHAPTER

3

NOSTALGIC SCENES

'We are going two hundred years back', announces a member of the theatre association *Jungfrau Theatre*, a well-known hotelier, as he introduces the historical play *Alpine Glow* (*Alpenglühen*) to tonight's audience.[1] 'On the valley, with the valley, for the valley', our play brings you closer to the valley and its people. And now, curtain up!' A crowd of villagers in traditional costumes from two centuries ago join him. Young girls walk barefoot, older women hide their hands under their aprons. Men and boys wear black embroidered jackets.

Thomas is among them. He has been recruited to play his 'own' role tonight as a genuine native thanks to his mastery of the disappearing local dialect. Someone next to him addresses both the audience within and outside of the scene: 'Dear people of the whole valley community (*Talschaft*) and from the area [he turns to the audience], welcome to the *Schwinget* [a traditional feast] on the Alp . . . For once, the patricians (*die gnädigen Herren*) from Berne and Interlaken won't mess up our lives!' Thomas answers, 'They are not strong enough to climb up to us in the mountains! They are too lazy (*faul*) . . . and only set the tone behind thick walls (*sie geben den Ton an hinter dicken Mauern*)'. 'But here, *here*', he insists, pointing to the ground, 'we set the tone!' The crowd cheers. A famous guest interrupts the party, coming directly from Berlin. It is the musician and composer Felix Mendelssohn, who declaims his awe for the mountains and their inhabitants.

Three scenes later, a last, grim vision of the village, set in 2030, appears before the audience. The valley is endangered by endless growth, urbanization and touristic development fostered by greedy *Touristikers*. The mayor of the future, a woman – an unprecedented event in the village's history – looks at a team of tourism lobbyists from the 'outside' as they present plans for future, exponential growth. 'What, even *more* guests? This means new infrastructure!' she exclaims worryingly. The team responds enthusiastically: 'More, more, more! The competition never sleeps'. 'But what about nature? We have farmers, we have Alpine farming, this needs to be preserved!' responds the mayor. 'We used to be the glacier [valley] (*Gletschertal*)'. 'Glaciers are dead, get used to it', the *Touristikers* cynically retort.

When I met Thomas at the beginning of 2020, over coffee he shared fond memories of the play with me. Like the mayor of the future, he said he made a point of still using the word '*Gletschertal*' to describe the valley where he was born more than fifty years ago, as a statement against a present he no longer understands. The last scene of the play was 'exaggerated' (*überspitzt*), but Thomas believed that this is what really could come upon the valley. The melting glacier, the overcrowded centre of the village, the digitalization of processes that used to provide work, the increasing speed with which people travelled the world, the urbanization, the centralization of administrative matters and the disregard for traditions, customs and values inherited from the past all felt profoundly worrying, he told me. 'But they are unstoppable (*unaufhaltbar*). That's why I paint the beauty of the valley – or what's left of it.' Nature – birds, sunsets, peaks – were some of his favourite motifs. Regularly, Thomas took advantage of his early retirement, waking up before dawn and driving to the beautiful viewpoints of the valley before the 'Asians' did. Sometimes he painted; sometimes he just enjoyed the calm and the beauty.

Looking at the crowds outside the café where we were, Thomas wondered how many more tourists could really be coming to his village. None of what he witnessed in the centre seemed sustainable, and the relationship between the lack of money for upkeep (as evidenced by the crumbling hotels) and the growth promised, imposed even, by companies like JR, were incomprehensible to him. Thomas concluded: 'But we voted [for the cable car], we wanted this development and now we can only search our own consciences. I also voted for the new cable car for the youth, our youth, to enable them to live in the valley – but now I regret it. I know many in the village do'.

Participating in the play *Alpine Glow* felt like a welcome escape from this difficult present for Thomas. It was not his first acting role either. After the golden age of tourism until the 1980s, when the global conjuncture felt increasingly uncertain and impacted on his ski rental business, Thomas discovered acting in the nearby open-air museum, where he featured in multiple plays written by Swiss author Jeremias Gotthelf, who described the impacts of modernization processes on traditional farming societies in the nineteenth century. When he heard that hoteliers wished to create a theatre association in his native village, he was enthusiastic, especially as he, like many other locals, had witnessed the shrinking of an already small culture scene. With everything dedicated to sports and tourism, events like going to the cinema and theatre plays were increasingly limited – another example of how tourism's predatory logics were dominating spaces and social activities (Azcárate 2020).

In contrast with many sceptical *Einheimische* like Moritz, who saw in hoteliers the threatening prospect of the *Touristikers*, Thomas was more sympathetic towards some of the local hotel owners. They had become important actors in trying to maintain a natural, pure, authentic image of the valley that also spoke to him. In a touristic region dominated by nostalgic touristic imaginaries, this was, at least partly, motivated by financial interests, but that did not make such involvement less important. Incentives to preserve the mountain village character, or *Dorfgeist*, of the place were welcome for Thomas, especially in times of unstoppable acceleration and growth. He knew all too well that hotels and hoteliers were part of the history of the place too, as some of his distant family members and friends had been involved in the hotel business until a couple of years before.

There was indeed nothing too surprising in seeing hoteliers display the past of the valley, in the play and beyond. Pictures of poor Alpine peasants and Alpine huts decorated hotel lobbies, an unavoidable part of the *Heidiland* or the ideal world (*heile Welt*) that the guests expected to see when they came to the Alps. As Adrian told me, the lobby of the ruined hotel had been decorated to please them and their tourist gaze (Urry and Larsen 2011). But there was something unexpectedly daring about the theatre play, at least to me. In contrast to JR's video 'Jungfrau Region 2026', where an *Einheimische* looks at a valley in ruins as a consequence of *not* innovating, the play precisely addressed how *too much* innovation and growth would destroy the valley, its peculiarity and heritage. I had regularly heard this critique from opponents of the cable car, described as 'backward' by many hoteliers, but when I questioned them, the latter insisted that the play was not meant to be read this way. In our conversations, the hoteliers who staged the play insisted that evoking history and the future creatively was part of a multifaceted endeavour. *Alpine Glow* was aimed at giving something 'back' to the local population in a village overwhelmed by mass tourism; something for them to celebrate and learn about their history and address the now trendy topic of climate change and the future in the Alps, but also to 'bring back' European, individual guests in an economy dominated by Asian group tourism. Around evocations of important moments of the local history and the future, the play was to gather hosts and guests in the present.

In this chapter, I will examine the critique proposed in this theatre play and address its stance on the touristic development of the region, along with its call for the valley's preservation and its goal to foster particular types of tourism. Whereas the previous chapter was focused on the creation of the future, this one will shed light on the omnipresence of the past in tourism imaginaries and reactions to tourism development, for both those who visit places and those who inhabit them. This partly speaks to the antimodern attitude mobilized by people like Moritz, who saw in the valley's rebellious, peasant past an inspiration for the present (see Chapter 1). Yet, Moritz's commitment to the native's freedom was undoubtedly motivated by his radical personality, as someone who was, in his own words, 'unafraid to speak his mind', and who even seemed to enjoy conflict and open quarrels with the *Touristikers*. He clearly could afford such a confrontational attitude: contrary to many of my interlocutors, none of his family members were employees of JR, he was not a hotelier, he had no children and he even seemed to benefit from a particularly masculine, independent posture that many saw as features of a noble saviour of the valley. Even though his family had been materially impacted by the cable car and the controversy that preceded it, Moritz told me that he had 'closed off' the topic and accepted JR's victory, and that this was a strength of character that some other JR opponents did not possess.

The characters who appear in the following pages are different. Although many shared Moritz's nativist vision, they were unable and/or unwilling to put it forward with such a forceful and radical tone. Because of their positions or personalities, the locals around which this chapter is centred mobilized a nostalgic, preservationist discourse that was critical to, but also compatible with, their activities and close ties with JR. Because of familial situations, their business activities, their networks or inclinations, hoteliers – but also locals like Thomas – were both seduced and worried about the *Touristikers*' power over the valley. Around the play, I examine their complicated position and the ambiguous character of their critique of the present, which appears both as a form of resistance to global trends in tourism and as a tool of touristic commodification, typical of Romantic imaginaries of the Alps. Nostalgic ideals of the pure, rooted life of native mountain dwellers once again appear here as integral parts of the village's reality – not only politically, as I demonstrated in Chapter 1, but also in the way one *does* tourism as a business.

A PLAY FOR THE VALLEY, WITH THE VALLEY AND ABOUT THE VALLEY

In a resort so oriented towards the needs and movement of guests, the theatre play's resolutely local focus, 'for the valley, with the valley and about the valley', already suggested a critical evaluation of the present. The play emerged in a context that Thomas and various locals felt deeply sceptical of, as they shared their concerns over the endless, or even exponential, touristification of the valley in the last decades. Over the years of my research, locals even gradually began to speak of 'overtourism' by using the English word. They echoed the concerns of other destinations globally, like Barcelona, Venice and Amsterdam, all faced with the impossible question of how to find a balance between touristic success, materialized in crowds and infrastructural sprawl, and a quality of life for those living off tourism (Colomb and Novy 2017; Cheer, Milano and Novelli 2019; Milano, Novelli and Cheer 2019). The democratization of tourism, the acceleration of means of transport, the change in consumption patterns and the gentrification brought about by tourism in the valley and beyond all underscored the latest phase in tourism's classical problematization or double-bind, as an industry destroying the pure views and experiences it commodifies (Enzensberger 1996). All the locals I talked to expressed how the problem had reached unprecedented levels with the recent emergence of 'mass tourism' in the resort, mirrored elsewhere in the country.[2] Echoing long-standing concerns over the matter in academia (Stronza 2005; Duffy 2006; Bianchi 2018), locals described this type of organized, industrialized and large-scale form of tourism as a new disruptive force – environmentally, economically and socially. In contrast with tourism scholars who have traced the origins of mass tourism to the Grand Tours (Chambers 1997: 12–15) or the post-Second World War period (Neveling and Wergin 2009), my local interlocutors situated the shift more recently.

For Thomas, but also Heinz, the JR executive I presented in Chapter 2, as well as many other villagers, the end of the 1980s marked a new era of intensive touristification with the arrival of tourist groups from the United States and Japan, and more recently from Korea, China and India. The latter travelled to Europe in no time in groups, transported by giant colourful coaches that filled the valley's parking lots and obstructed its (main) roads and sidewalks. 'Mass tourists' stayed for anything between a couple of hours and a 'maximum' of two nights, as Heinz said. They were mostly interested in visiting the sightseeing platforms owned and advertised by JR throughout the world, notably via Frei's marketing trips in Asia. As I discussed in the previous chapter, Heinz even explained that JR had been able to grow precisely because of these tourists and their greater predictability. Out of the one million guests visiting JR's infrastructure during my first stay in the region in 2017, about 70 per cent were Asian and mostly on organized trips (Seiler 2016). Thirty years prior, JR had registered only 280,000 visits, with an already important but fairly insignificant number of Japanese tourists compared to the 2010s (ibid.). For the village specifically, however, the exact number of such day trippers remained 'incalculable' as one tourism lobbyist explained to me: because of the diversity of transportation means (train, bus, car), as well as the fleeting presence of these guests, they did not appear in the local 'overnight stays' statistics. If the resort had, technically speaking, welcomed its first British 'group tourists' of the Thomas Cook tours at the end of the nineteenth century, there were undeniable indications that group tourism from increasingly far away had gained in importance. Yet, the exact contours of such reality remained vague – another manifestation of tourism's omnipresence and uncertainty.

If belonging to the 'toured ones' (*Bereiste*) generally meant to live a life dominated by the needs and presence of 'travellers' (*Reisende*), all locals described the newly brought-in masses from far away as *particularly* difficult to reckon with. Mass tourism stood for a fast, crisis-sensitive and dissatisfying form of tourism. As one of my interlocutors once expressed, these guests were the type of sheep-like consumers of the twenty-first century: unaware, uninterested and alienated from genuine values and processes, which was most visible in their love of ready-made goods and experiences. One group of critical locals (albeit not all based in the valley) expressed in a manifesto that JR pursued 'quality instead of quantity' (*Qualität statt Quantität*) – a phrase repeated by countless villagers. As this group explained, they opposed the company's project not primarily out of personal concerns but because of the very ideology standing behind it: the new cable car was to 'accelerate mass tourism, according to the motto, "Up quickly and away again quickly"'. The new cable car, one could read in their statement, would transform the '[*Dreigestirn* mountains] into a fast-food experience'. They rejected the idea that a trip to the mountains should be made into a 'commuter route' (*Pendlerstrecke*) based on 'speed and efficiency', instead of enjoyment of the 'dramaturgy' offered by the meandering train ride to the tops.

The environmentalist NGOs who opposed the cable car because of its impact on a landscape of 'national importance' (see Chapter 2) presented arguments that, if less provocative, went in a similar direction.[3] The link between a 'massive' project and its main users, 'mass tourists', according to many in the valley, was easy to make. Perhaps more surprisingly, even supporters of the railway company did not hide their dislike of mass tourism. Anna, the young JR employee and fervent supporter of the company, echoed the sentiments of many of her friends and relatives when she highlighted the paradoxical effect of 'mass tourism' – which she, like everyone in the valley, used as a synonym for 'group tourism'. Mass tourists, she contended, simultaneously took up too much space as they concentrated in hotspots, and so only passed through the village. Yet, Anna said, accepting the presence of these guests was part of the agreement or compromise with the company in exchange for a future in the valley.

In this context of scepticism towards mass tourism, incentives began to emerge aimed at selling the beauty and traditions of the valley. One loosely organized group of more or less active hoteliers, accompanied by a handful of tourism marketing coaches, were involved in the selling of products of 'good', 'unique' and 'original' quality. These were exclusively produced within the borders of the municipality, as has been the case in many Alpine valleys selling heritage products such as food and artefacts (Grasseni 2016; Sidali and de Obeso 2017; Lems 2023). Otherwise omnipresent in the regional touristic landscape, JR was not really involved in their activities, as one of the group's founders explained when I enquired about the company's participation in an interview. For many in the valley, the group was pushing for something different from dominant forms of local tourism development, namely a form of tourism more respectful of the locals, traditions and nature. So different that many locals predicted this initiative 'would not last' in an environment where bakers, craftsmen and butchers had disappeared one after the other. Upon meeting with interested locals like Thomas, the hoteliers came up with the idea of a theatre play in 2017.

Alpine Glow was the group of hoteliers' most audacious project, they explained during the multiple public and private information meetings on the play. The team of hoteliers and their partners mobilized local slang, particularly when one of them referred to many locals' weariness of the fleeting, anonymous and 'excessive' (*masslose*) presence of 'Asians walking with their plastic bags in their hands'. They joined in a very widespread discourse in the valley, criticizing mass tourists for their consumption of places and goods by visiting malls, buying noodle cups or frenetically taking pictures with their phones, instead of enjoying homemade Swiss food, consuming it in local outlets, and enjoying the moment. The play's team referred to a common local narrative that emphasized that Asian mass tourists in particular did not *really* spend money locally, even if they were physically (but shortly) present in the resort. According to it, what and how these guests consumed – mostly imported and packaged goods, from big companies that had their headquarters outside of the valley – revealed that they were not truly invested in connecting with the 'real' village and its traditional, family-owned shops and restaurants.

The play was to respond to this situation by doing something for the locals and bringing local life back to the village centre. The hoteliers' team stated that the play was an initiative *for* the local community, as well as to celebrate the traditional local clubs and display a local history that was to be instructive. At the same time, *Alpine Glow* was intended as a business strategy to attract those European but mostly Swiss guests who would be interested in history and the play's message. The play would be a 'reminder' that the resort exists for the Swiss and this would incite them to come 'back' to the village as tourists. By saying this, the hoteliers responded to the concerns of villagers witnessing the number of European travellers receding following the various financial crises and the strength of the Swiss franc. The Swiss represented a rather stable and important, albeit slightly receding, segment of the market. In contrast, the non-European and especially Asian market had grown significantly.[4]

Historical accuracy and authenticity were at the centre of the team's artistic choices: the theatre scenes had to 'feel real' by having people speak in their native mother tongues, they said. The *Einheimische* locals would play their own roles and speak the local dialect, while the English characters of the play would speak British English and the German characters standard German. The play could be partly understood by non-dialect German speakers thanks to the 'dramatic power' of the play and the music, but such folk were not the intended audience. 'We want Swiss-German tourists from around here, from the canton of Berne, locals (*Einheimische*) if you wish', one of the organizers explained to me, mobilizing a much more flexible understanding of who was deemed local (*Einheimische*) than the nativist one shaping local politics that I described in Chapter 1.

The motif of 'glowing' mountains was the central idea of the play, one of the organizers and main initiators of the play told in various meetings. It was to thematize the fascination that the region had exerted on tourists over the centuries, with its magnificent mountains drawing in guests from afar. The play also addressed the problematic future of the valley and beyond in the age of climate change by putting the spotlight on glaciers that had been 'literally glowing' since the they had first formed, yet were now slowly but surely disappearing from what had once been the 'glacier valley'. However, one of the play's organizers insisted on the importance of not being 'moralizing' in such issues: 'I don't like to tell people what to think', he said to me. Instead, the play was to present what the local youth would like to have for a future. Targeting locals and guests, and addressing the past and the future, the theatre project was also envisaged as a sensorial, total experience, set on an open-air stage to take advantage of the view of the real mountainous landscapes: the *Dreigestirn* mountains, the (remaining) glaciers and the sunset.

After the recruitment of local actors, and now with a team of approximately one hundred participants between the organizers, actors and helpers, the play was shown to the public for a month in summer 2019. Just about every 14 show was sold out, and many of the people I knew in the village enjoyed it: they had learnt something, had laughed, or had been touched by the final scene of the ruined future. All historical scenes were well-known episodes of the local history 'from farming to tourism' that they recognized or were reminded of. The futural scene gave food for thought, as some locals stated, echoing the very positive newspaper reviews. 'It was a tremendous and unanimous success', Thomas summarized: he had been congratulated on the streets, and the municipal authorities had officially thanked the cast for their efforts. Yet, this was a one-time event and was not meant to be repeated. It cost too much in time and effort, and in itself was not a sustainable activity – Thomas knew that. A couple of months later, another local hoteliers' initiative of selling traditional products was also in difficulty, on account of unfair competition from bigger players.

Such local efforts to promote and sell other goods, to push for another type of tourism in a context of acceleration, may seem marginal, perhaps even more so given that these initiatives were short-lived. Yet, they speak of a global trend of differentiation in tourism development: since the 1980s and 1990s, alternative forms of tourism such as ecotourism, and more recently 'ethical' (Gibson 2010) or 'humanitarian' tourism (Mostafanezhad 2014) has emerged in response to the always more tangible detrimental natural and social consequences of 'mass tourism' (Duffy 2006). These developments have aimed at giving 'back' to local communities and educating tourists, both, albeit differently, disappointed by the dominant forms of tourism development. However, as scholars working on the topic have shown, such initiatives have often reproduced the unsustainable logics they denounced in the first place (Azcárate 2020: 72). In the touristic European Alps, this trend has taken the shape of what can be defined as 'slow tourism', a more sustainable model of tourism contrasting with dominant and 'fast' modes of food and travel consumption (Heitmann, Robinson and Povey 2011; Grasseni 2016; Sidali and de Obeso 2017; Lems 2023).

At the same time, the play and its instigators reveal something of a reality that extends beyond the realm of tourism: what sociologist Hartmut Rosa has described as an important, vibrant culture of the critique of acceleration. As he points out, 'the call for deceleration and the nostalgic desire for the lost "slow world"' is culturally hegemonic, in spite of its coexistence with a reality of the 'victory of the accelerators, i.e. with the introduction and entrenchment of the new technology' (Rosa 2013: 41). Delving into the play *Alpine Glow*, I will examine how such decelerationist critique indeed appears to be profoundly nostalgic and culturally dominant in a region so historically and economically attached to Romantic imaginaries of timelessness and authenticity.

ECHOES OF AN IRRESISTIBLE CALL

When I met Thomas, I understood why he was cast to play his own role, as a native peasant from two centuries ago. Although not a farmer (or, rather, not anymore – a part of his family used to own a farm), he was one of the few natives who still mastered the dialect and with that, he served as a living example of the accuracy and authenticity sought by the play organizers. Thomas also helped to write dialogues in local dialect when outsiders were to play natives, as not enough 'genuine' actors could be hired to play their own parts. Yet, upon looking at Thomas's paintings and the energy he put into representing an intact, beautiful and natural valley, I could not help but notice how he also resembled the figure of an outsider: Mendelssohn.

In the first scene of the play, Felix Mendelssohn explains to the crowd of natives that he is irresistibly attracted by the mountains, and that they had drawn him, all the way from Berlin. He is accompanied by elegant English travellers, served by domestics. The natives gather around these strangers, fascinated, as the tourists turn to the crowd and say: 'Look at these marvellous natives! And the mountains!' Mendelssohn expresses his fascination for the glaciers, to which the natives respond that they, the glaciers, are a threat to their existence, as they grow and 'eat their land', which could potentially force them to leave the valley. 'But they attract foreigners like me, and guests bring money to the valley', responds Mendelssohn. The villagers shake their heads and explain: they are (still) poor farmers who own only a small plot of land with a cow or two and a goat. 'There is nothing to admire here', they conclude. 'Are you blind?' exclaims Mendelssohn, outraged, turning around and pointing at the peaks surrounding the open-air stage: 'The Alpine glow calls us (*ruft uns*) . . . Who can resist it (*widerstehen*?)'. 'For us it's the everyday (*alltäglich*), we belong to them and they belong to us', says a native woman, shrugging her shoulders. 'How do you do this Alpine glow?' asks Mendelssohn jokingly. The natives, who do not understand Mendelssohn's ironic sense of humour, further explain in a local dialect that Mendelssohn is struggling to understand: the Alpine glow is the result of the local mountain's spirit work as he alights the peaks with a torch. Moments later, the villagers sing a yodel song and Mendelssohn listens to it admiringly, seated in front of them. The composer's String Symphony No. 11 in F Major, which he wrote after his journeys in the Bernese Highlands and his encounter with yodel, is played.[5]

When I met with the one of the play's creative assistants for an interview, she described this scene as one of first contact between the worlds of the toured and the travelling clashing – between poor, unaware, naive natives and rich, educated outsiders. Such a scene had great potential, she said, as it depicted a powerful historical reality that could be used for its comical character. The assistant was right on these two accounts: presented with the unaware native's naive remarks and the sophisticated taste and visions of travellers, the audience regularly burst into laughter during this first scene, which was packed with humorous quid pro quos. At the same time, there was indeed a historical basis to the scene. Romantic artists and intellectuals, like Rousseau and, later, Mendelssohn, as well as the wealthy guests they then inspired, were urban elite members who travelled to an Alpine world of poverty and difficult living conditions.

Thanks to their greater economic, social and cultural capital, these outsiders imposed their ideal and (literally) romanticizing visions of the simple mountain life to satisfy their own quest for meaning, inner truth, expression and authenticity in an alienating world (Berlin 1976; Stone 2014) – before leaving again to present their creations to their wealthy, educated and connected urban networks. Mendelssohn composed pieces, wrote poems and drew sketches inspired by no less than four trips to the Swiss Alps, from the age of thirteen (Gallusser 2021). But what the lives of highlanders actually entailed beyond these visions has remained largely unexplored, as historians of the Alps themselves have argued (Mathieu and Boscani Leoni 2005).

As clear as the gap between the mountain dwellers and the Romantics had been, the play seemed to reveal its pastness. Two centuries after this foundational moment for the history of the valley and modern tourism, the Romantic vision had clearly won: paintings, postcards, music, books and houses portraying the authentic, rural life in the mountains saturated the village's visual culture. Instead of seeing nature and mountains as posing a danger to human life – a common perception of the Alps before the Romantics (Mathieu and Boscani Leoni 2005) – local participants like Thomas and the play's organizers were now convinced of the opposite. The increasing technological development and connection stood for the disappearance of a grand nature, made up of authenticity, purity, and meaningful relationships. In a world of 'unstoppable' acceleration, Thomas believed that the valley 'still' offered a slice of beauty and exceptionality, a safe haven from the rest of the world, in the same way Romantics like Mendelssohn had turned to the mountains as a site of yearning when they witnessed the rapid industrialization of their urban lowlands.[6] For Thomas and the play's organizers in the twenty-first century, and for Mendelssohn in the nineteenth century, what was left of the glacier valley needed to be celebrated and preserved, no matter what. The play not only thematized the encounter between lowlanders and highlanders but also subscribed to the Romantic ideals of the former.

This echoes with tourism sociologist Adrian Franklin's description of touristification processes that turn the ordinary, banal and scary into extraordinary, beautiful and unique experiences (Franklin 2008). As a form of 'ordering' the world (Franklin 2008), tourism shapes the expectations and perceptions of travellers (Urry and Larsen 2011), but it also influences the world of the 'toured ones', who enter into new frames of interpretation and perception of their own home environments. This tourism ordering is of a spatial nature: tourists travel to destinations in expectation of experiences linked to a special natural and cultural environment, and locals in turn begin to ascribe new meanings to their villages as they become touristic – an effect that anthropologists of tourism have observed in certain destinations (e.g. Tucker 2005; Leite and Graburn 2009; Chio 2014). But following the vision of Romantics like Mendelssohn and of locals like Thomas, tourism orderings also, importantly, appear to operate on a temporal level, according to nostalgic principles that have their roots in the Swiss Alps.

THE SWISS DISEASE AND THE ALPINE ANTIDOTE

In his dissertation published in 1688, Doctor Johannes Hofer (1934) wrote about Swiss soldiers who had mysteriously deserted, fallen ill or died on the battlefields of Europe. He claimed that these young mercenaries, many of whom were peasants and had migrated far from their own countries, suffered from a medical condition he called 'nostalgia'. While this phenomenon was known previously as *Heimweh* or *mal du pays*, Hofer was the first to pathologize it as a disease (Starobinski 1966: 84; Lems 2016: 3). Born out of the contraction of two Greek words, *nostos* (the return to the homeland) and *algos* (suffering or grief), the word 'nostalgia' was Hofer's diagnosis for the mercenaries' medical condition, 'a state of fever and lassitude that was potentially fatal if left untreated, but readily cured by a return home' (Atia and Davies 2010: 182). The young Swiss men, away from their mountain valleys and exposed to 'foreign air' (Hofer 1934; Starobinski 1966: 88), were to return where they naturally belonged: in their valleys, their mountains, their *Heimat*. In a later version of his argument, Hofer explained that the disease of nostalgia was typically induced by a song titled '*Kuhreihen*' or '*Ranz des Vaches*', which deeply touched the soldiers in their hearts and bodies. The song recalled the mercenaries of places and times left behind, of their lives as peasants calling their cattle while going up or down the Alp (Lems 2016: 3). Following Hofer's findings, nostalgia was soon to be characterized as the 'Swiss disease' (*Schweizerkrankheit*).

By the end of the eighteenth century, the evocative power of the *Kuhreihen* motive propelled the Swiss disease into the realm of philosophy and literature. The tune was incorporated into works of Romantic artists, musicians and writers. Rousseau, Schiller, Goethe and Mendelssohn[7] were fascinated by the figure of the longing Swiss and his despair upon hearing the song

(Atia and Davies 2010: 182). Nostalgia was no longer a negative, debilitating illness but rather the symptom of a sane realization, vested with symbolic and metaphorical power (Casey 1987: 363–64; Salmose 2019). Contrary to Hofer's medical theory, the Romantics did not believe in a curative return to the homeland for those suffering from nostalgia: like all those who lived in an increasingly meaningless world of progress, they were doomed to long for objects that times and society had irremediably transformed (Jankélévitch 1983). The Romantics understood nostalgia as 'the metaphorical expression of a much more profound rupture, the separation of man from ideal' (Starobinski 1966: 95). From a sense of geographical disruption, the Romantics contributed to the shaping of nostalgia as the temporal concept and affect we now know, namely what one feels for an object or a time that is disappearing.

In the widely circulated nostalgic portrayals of Romantics, the Alps and the life they promised emerged as bulwarks against debilitating processes that were transforming Europe at its core (Denning 2014). Industrialized at a later stage than the lowlands, these mountains were the place of harmonious and organic coexistence between humans and nature – where natives were 'at one' with the environment, as the villagers of the play insisted (see also Schär 2012, 2015). Travelling to the Alps became a temporal and moral endeavour, back to the origins of mankind and civilization even, as the unique vestiges of a time of authentic values. The Romantics' fascination for the Alps became so dominant in the Western world that it also gave rise to countless initiatives to preserve and valorize the integrity of Alpine regions. Included in national conservation schemes from the end of the nineteenth century, the Alpine ridges stood for precious symbols and monuments of the Swiss *Heimat* and its origins (Zimmer 1998). The integrity of the Swiss nature became a matter of national heritage (*Heimatschutz*), leading to the emergence of regulations and inventories aimed at safeguarding the environment in the twentieth century, such as the Swiss Federal Inventory of Landscapes and Natural Monuments (Summermatter 2010; Bachmann 2012).

In 1984, the Jungfrau-Aletsch region, where my field site lies, was inscribed in this national inventory (precisely what enabled the environmental NGOs to appeal against JR's cable car project, as described in Chapter 2). The Romantics' legacy also contributed to the inscription of the Jungfrau Region as a UNESCO World Heritage Site in 2001, at a transnational level.[8] The reasoning behind this inscription still lies in the exceptional character of the region, spared by human developments, as one can read on the Jungfrau-Aletsch website: 'The Swiss Alps Jungfrau-Aletsch UNESCO World Heritage Site is one of the few areas in the Alps with only minimal anthropogenic influence [and] . . . intact nature, surrounded by settlements'.[9] Consecrated as objects of universal care and attention for their natural and cultural character, the valley and the Swiss Alps were to be preserved in the name of 'global cultural commons' (Geismar 2015: 76).

In his work on the heritage site of Luang Prabang in Laos, David Berliner (2012, 2014) notes that heritage practices result from intrinsically nostalgic efforts to preserve objects over the irremediable passage of time. Heritage nostalgia, he states, glorifies the irretrievable past 'without necessarily implying the experience of first-hand nostalgic memories' (2012: 770).[10] If based on 'discourses of loss detached from the direct experience of losing something personal', this form of nostalgia is no less of an emotional one, as it may involve 'indignation, anger or pain' (Berliner 2014: 21). While Berliner bases his argument on heritage, such conclusions also apply to many touristic destinations, especially rural, which operate on principles of uniqueness, purity and timelessness in a landscape of homogeneity, degradation and speed. In the Alps and in many other institutionally recognized heritage sites, the realms of the touristic and the heritage intersect, as they consecrate sites as exceptional and worthy of conservation, admiration and visit.

Thomas's own wishes to see the re-emergence of more authenticity not only resemble Mendelssohn and his peers' yearning for the non-alienated world; they are also products of this way of comprehending the world and its transformations. These sensibilities are the result of the democratization of nostalgia, from a personal experience of loss and exile endured by Swiss mercenaries into a second-hand experience or 'exo-nostalgia', in Berliner's words (Berliner 2014), experienced by somewhat privileged travellers, institutional actors, environmentalists and locals. As much as my village interlocutors expressed their discomfort with what they perceived as an external pressure to subscribe to the image of a timeless (and often entrapping) Romantic Alpine village, many, like Thomas and Moritz, often performed these ideals for themselves (Leitenberg 2022). They did not do so solely to conform with the travellers' expectations or 'tourist gaze' (Urry and Larsen 2011), but also because they were themselves affectively invested in these ideals, as Jenny Chio, for instance, also noted in her work on ethnic tourism destinations in China (Chio 2014). The presence of tourists from all over the world and the institutional making of the Alps as heritage sites, if problematic in some respects, were confirmations of the exceptionality the valley, its people and its culture for many locals. Having shown the replications of nostalgia as a historical concept and a sensibility that has traversed the centuries between insiders and outsiders, I now come to the last, most surprising scene of the play – one that appeared to be, at least at first, far from any romanticized images of the Swiss Alps.

THE BLUES OF THE FUTURE

In the play's fifth and last scene[11] set in 2030, the mountain spirit, comes down from his ice cave before the debate between the mayor and the *Touristikers* with which I introduced this chapter. The spirit tells the audience that he might have to leave his home valley because of what we understand to be global warming. After the spirit's nostalgic complaint, the *Touristikers* gather in front of a travel agency called the 'New Horizon Agency'. The most vocal *Touristiker* speaks in empty slogans and draws on exponential graphics showing profit margins of all kinds, derived from the financial exploitation of the 'optimized Alpine world (*Alpenwelt*), which is budgetable, predictable and under control'. He complains about nature and cows, which stink and roam freely and are also 'guilty of climate change'. 'Climate change? Fake news!' his team responds. 'We want expansion, explosion . . . and we need to react, reorganize, reoptimize . . . and make profit! For the guests, only the best!' The play's band starts playing a song that denounces the 'building mania' (*Bauwut*), the 'competition' (*Konkurrenz*) that divides the valley, and the disruption caused by new hotels and infrastructure. At the end of the tune, the mayor, who is being proved right by the song, invites young people 'who want to live in the valley after us' to speak.

Half a dozen youngsters enter the stage and stand in line in front of the audience, without costumes. 'Have you heard this? We don't want that!' says a young boy. One after the other, they express their wishes to 'inherit something, something meaningful', something 'nice', nature and perspectives:

> Glaciers are not snow from yesterday, they show the valley what happens in the world. The stories of the past are not enough, they are postcards from a time we no longer know . . . You think we were influenced by Greta [Thunberg], that it's a trend and will pass. It's wrong! We want to have something to be proud of, to show to the guests.

The show ends with a new version of the local anthem, calling for the preservation of the village where it is 'nice to live'.

Like the Romantics who saw the *Kuhreihen* as a critical symbol in their time, the last scene of the play points out the critical potential of a nostalgic, preservationist discourse. Time's passage and progress are irreversible, a source of pain for those in the play, like the youngsters who wish to 'keep' the glaciers, the integrity of the beautiful valley, farming, skiing – and even their pride. As much as the Romantics' nostalgia targeted notions of economic and social progress, universalism and rationalization in their time, the play points to a more contemporary use of nostalgia that still allows for the critical evaluation of the present. In postmodernity, many have even used nostalgia to symbolize what it means to live and to react to a world of unstoppable acceleration and fragmentation where loss or exile have become the new normal (Jankélévitch 1983; Angé and Berliner 2014). In historian Peter Fritzsche's words, yearning for irreversible, golden times is 'a dramatization of discontinuity' (2001: 1610) that reveals the spleen of the alienated postmodern subject and the feeling of human decline accompanying technological progress. Nostalgia in this context is a critical and palliative tool, sometimes capable of empowering subalterns or opening the breach to alternative modes of existence as it sheds light on the problematic state of the present, and potentially opens up utopian futures (Stewart 1988; Hodges 2010).

Although the organizers of *Alpine Glow* explained that they refused to be 'moralizing', the play literally dramatized discontinuity by making strong claims about the futural consequences of the endless development of tourism under the impulse of *Touristikers*. The 'exaggerated' vision presented in the play, by Thomas's own accounts, found echo among many locals because it responded to pre-existing experiences of longing for times of greater ecological stability, such as the visual presence of the glaciers' blue ice – now disappearing. The youngsters' plea in *Alpine Glow* to act 'before it is too late' played on feelings of (future) nostalgia but also regret and responsibility over what could have been done, in a way that eerily echoes the future anterior of JR's video, which I described in the previous chapter. At the same time, the youth's call for continuity and pride cannot be divorced from a global reality of environmental destruction, increasingly met with what Angé and Berliner have recently termed 'ecological nostalgia', consisting of 'attachments to living organisms that are jeopardized, or already gone' (Angé and Berliner 2021: 9). In rural Alpine destinations visited primarily for their natural beauty, as well as being under strong pressure from global warming, Herta Nöbauer has revealed what such ecological nostalgia implied for her local interlocutors seeing their 'existence literally melting away' with the melting of glaciers (Nöbauer 2022). In the Bernese valley too, the play revealed an ecological nostalgia highlighting the multiple forms of natural, sociocultural, economic and affective heritage that the Alps represent: an environment, a capital, a culture, a symbol.

BETWEEN CRITIQUE AND COMMODIFICATION

Thomas saw in the play a representation of his own regrets of having voted for the cable car. The play symbolized that valley dwellers who, like him, had voted in favour of the new infrastructural project now had to 'search their conscience', as he put it. His assessment of the valley's latest development however contrasted with the cautious position of the play's organizers, unwilling to be too critical of the *Touristikers* in the tense context surrounding the new cable car.

Attempts to temper the play's message, were indeed spread throughout the scenario, apparent for instance in the final scene's unexpected direct address to the audience by the nostalgic mayor. After vehemently opposing the absurd plans of growth and commodification imposed by the *Touristikers* – by, for instance, getting rid of cows entirely – she broke the fourth wall with a cryptic message to the audience: 'one can be for or against it [the *Touristikers*' plans]'. The organizing hoteliers also claimed that the play had 'nothing to do' with the cable car controversy. The fact that the *Touristikers* of the play spoke in slogans that were not dissimilar to those which tourism lobbyists mobilized in the present (see Chapter 2) was just a matter of interpretation – my interpretation.

The critique presented in the play was also carefully tailored to speak to local political inclinations.[12] While the play's message opposed the fictional *Touristikers*' denial of climate change, it was also critical of 'trendy', global ecological activists like Greta Thunberg, a figure despised by many locals. Many of the villagers I met dismissed these views as those of naive, morally superior and lazy young people. The latter gave farmers and rural dwellers lessons on how to live in accordance with nature while being alienated from it themselves, as cosmopolitan urban dwellers – a common critique of social movements based on populist and conservative discourses of 'common sense' (Wodak 2015; Lems 2022; Pasieka 2022). Being 'green' in the village was an insult, an indication of lack of moral integrity or compass. One popular saying originating from conservative circles in German-speaking Europe circulated on how the greens were like 'watermelons': green on the outside but red on the inside, and full of water – nonsensical liberal ideology. Many of my interlocutors responded to the environmentalists' concerns over Alpine development (which they saw as reproaches) that they already led much more sustainable lifestyles than others elsewhere. They perceived themselves as stripped of alternative means of subsistence outside of tourism, in an entrapping leisure area created for longing urban dwellers in the first place. Villagers thereby mobilized what geographer Mathieu Petite has described as 'localist' discourses, emphasizing a mountain dweller perspective on the Swiss Alps as a place whose future they should determine instead of the 'eco-freaks' (Petite 2011, 2013).

It is this political and economic background of the village and the Swiss Alps that shaped the form and constraints of the critical potential of the play *Alpine Glow*. Not being 'moralizing' meant formulating a cautious critique that would not offend or confront the real life *Touristikers* and their supporters as crucial business partners in the region, as well as adapting it to the political sensibilities of rejection of global environmental activism. The play's critique was a particular, embedded, local one – for the valley, on the valley, by the valley.

But the genealogy of the Swiss disease also points to a degree of complacency, or even compatibility, of nostalgia as such towards the very forces it denounces. Although it implies unease or suffering, nostalgia has rarely resulted in overthrowing capitalist modernity. It rather tends to reassert its naturalness: the modern linear passage of time might be painful but it is also irremediable, unstoppable, irresistible. Moreover, since the Romantics, the emotional longing for lost places, times and objects has played a crucial role in capitalist forms of commodification (Jameson 1991; Robertson 1992: 158). This is true of tourism in particular, and even more so of natural or cultural tourism – both of which the valley embodies – where imaginaries of authenticity derive from a nostalgic posture towards the countryside, the landscape or heritage as markers of a greater, but lost, past (Graburn 1995; Urry and Larsen 2011; MacCannell 2013).

As an endeavour of hoteliers, the play did not escape such commodification logic. During meetings and events destined to inform and convince the local population to support the artistic project, the organizers regularly insisted on the value of European and Swiss guests. *Alpine Glow* served to attract such guests, 'reminding' them of the resort's existence but also raising its 'attractiveness' as a destination for the European tourism market. 'Selling history and stories about the village' directed at 'locals and tourists interested in history, the village and its traditions' was the marketing strategy, as one of the hoteliers exclaimed, echoing the alternative tourism strategies adopted by many of their peers in the industry (Stronza 2001). In facilities saturated with images evoking the simple, pure and peasant mountain life – hotels, congress halls, the village's streets – the hoteliers argued for yet another way to sell the village's uniqueness and authenticity in a world of homogeneity and speed. In this spirit, the theatre group refused to sell food and drinks but purportedly pushed people to consume on the *Hauptstrasse*, in some of the remaining local shops, and encouraged them to purchase packages in local partner hotels consisting of overnight stays with meals and theatre tickets.

Most Swiss and other Western tourists I met indeed expected the village to offer the nostalgic tourism that the hoteliers proposed, and they were ready to pay for it. The valley, they said, was a place of local traditions and rugged sights, contrary to 'overtouristified' destinations such as Zermatt or the town of Interlaken, where everything was designed for tourists. Many of these tourists purportedly avoided crowded, inauthentic spots. 'Here there are still cute, wooden chalets', one skier explained to me, favourably comparing the village against Davos and its more urban (and thus less Swiss) feel. When I met an American couple of Swiss extraction, from Berne, Indiana, they described a valley of environmental and cultural integrity, contrary to the problems their country was encountering with 'illegal migration and Islamization'. They told me they wished their great-grandfathers had not chosen to flee poverty in the Swiss Alps to emigrate to the United States, because life 'still seemed easy' in the Swiss Alps. The couple had travelled to the valley and spent significant amounts of money in order to witness and experience traditions that evoked a simple, genuine life. Even for those who, like them, shared an awareness of the problems their ancestors had encountered in rural Switzerland two centuries earlier, visiting the Alps seemed like a trip not only to connect with their roots but also to experience a nostalgic image of the region.

Between critique and commodification, the shape-shifting quality of Alpine nostalgia seemed to have something for everyone in the valley: locals, tourists, tourism 'makers'. But was there anyone it did not speak to, anyone it did not speak of? Discussions of global calls for a return to a nostalgic past have shown how these often entail a critique of social relations, excluding some for their ethnic, racial or religious background or gender identity (Betz and Johnson 2004). In the remainder of this chapter, I evoke how the village was not immune to this reality by discussing what or rather *whose* nostalgic visions of the Alps seemed to be evoked here – a reflection I will continue in chapters 4 and 5.

TIMES OF CREATION, TIMES OF DESTRUCTION

Between Mendelssohn's first scene, set in the 1830s, and the final scene representing the future of 2030, the play told the story of alpinism until the skiing boom that took off in Switzerland in the 1970s. Tourists from England, Germany, Italy and the USA, played by (mostly non-native, outsider) locals, were integral parts of scenes evoked with the same light, entertaining tone as the original scene of encounter. In the final dystopian scene, the tourist crowds were omnipresent, in the *Touristikers*' words, but absent. The 'business of foreigners' was displayed without foreigners, and the events unfolding between the second half of the twentieth century and a future of ruination in 2030 were truncated, but they seemed dramatic. The countless accounts of locals I met over my research allowed for an explanation that I doubted the tourist audience of the play could really understand. From an economy of stability and local profitability, as the local narrative had it, the mass tourism pursued by *Touristikers* made for an increasingly dissatisfactory village life and uncertain economy – a situation that Heinz, the JR executive, contended was 'challenging' for locals (see Chapter 2).

There was some truth to this nostalgic temporal reasoning, on a local and global level. Globally, the intensification of tourism dependency and the growing monopoly of a few players has contrasted with a more locally anchored, heterogeneous and less intensive tourism industry, before globalization (Bianchi 2018; see also Chapter 4). The past was also marked by a greater environmental integrity in the valley: the snow used to fall in huge quantities – enough to block the access to the valley in a memorable winter in the 1990s; the glaciers' blue ice used to be visible from all sides of the valley. With these processes, the locals' dependency on tourism had undeniably been transformed. The play portrayed this gap between new and old forms of tourism set in radical opposition, differentiating between an economy generative of positive forms of dependency, and an essentially destructive one for its dependents. But the absence of guests in the final scene unveiled a deeper layer of local perceptions regarding this transformation – one centred on the emergence of new visitors and their role in fostering undesirable forms of dependence on tourism.

THE NOSTALGIC AND THE NEW GUESTS

In autumn 2023, I was sitting next to Felix, a local postcard enthusiast, in his office when he pointed to a picture of tourists in fancy clothes sitting in the valley's main street during the Belle Époque. These guests, probably British, he said, were 'sophisticated' (*erhoben*) – symbols of the high time of tourism. These travellers belonged to the life of the village: they stayed for weeks, sometimes months. They built a church in the valley, and their hotel rooms were like second homes, tailored after their individual tastes, as they returned every year. He knew all about them because they occupied a prominent place in the history books on the valley, and even made for a touristic attraction: one of JR's installations was a 'nostalgia' train that allowed one to experience the tourism of the Belle Époque in the region. Alpine farming times had been essential to the creation of the valley as a touristic resort, but the history of tourism itself had now become part of the romanticizing tourism imaginaries.

Felix looked nostalgically at the postcard from a position of bitterness towards the present. Echoing the sentiments of others, he witnessed the phenomenon of Asian 'mass tourism' in the valley with great dismay. These guests had changed the face of the village, with new restaurants and shops responding to their needs now flourishing in the valley. He praised the British guests of a century earlier for creating and transforming the valley thanks to their presence, but with the 'mass tourists' it was different. Felix seemed to point to what Heinz, the JR executive, calmly but bluntly stated when I had met him during the first year of my research: '[tourists are] now differentiated between good and bad guests'. Heinz's company could draw on unthinkable successes thanks to the greater predictability of group tourists from Asia, but Heinz was aware that locals disliked mass tourists because of their 'non-profitability' and the difficulties inherent in welcoming guests from different cultural backgrounds, as opposed to the regular guests (*Stammgäste*).

Later on, I came across a comment posted on Facebook by a native villager, stating that his native ancestors had 'sold off a lot of land'. He continued:

> The village with the landmark [*Hausberg*, home mountain] is in danger of being lost with mass tourism. The *Dreigestirn* . . . serves mainly mass tourism, where most of the added value flows out to the valley. This strategy is wrong. The '*Dreigestirn*' stands more for mass tourism . . . but the real annual added value from tourism for the locals is based on individual tourism. . . . The [*Hausberg*] is and remains [the valley's] distinctive local mountain . . . and [stands] for high-quality individual tourism with privileges for customers. In terms of marketing strategy, the two tracks [of individual and mass tourism] must be clearly separated. For [the valley], the future lies in individual tourism with intact mountain farming.

Asked by a German tourist who had come to the valley multiple times to explain what he meant by 'mass tourism' in the comments, the author of the message described individual tourism as 'single travellers, families and small groups such as hiking groups or school trips'. 'Mass tourism is the big buses with quick stops here and there.' The German tourist, he implied, was one of the valuable individual guests.

The concerns of people like Felix, Thomas and this user often took the shape of a popular saying, both in the valley and the country, that I briefly mentioned in Chapter 1: the '*Ausverkauf der Heimat*' – an upsetting oxymoron, which stood for 'the selling of the homeland', the alienation of what should remain unalienable. Whereas the *Ausverkauf der Heimat* was a critique directed at the makers of (mass) tourism, it also targeted the new consumers of tourism as both economically unprofitable and culturally alien. New and disruptive, these guests were not only absent from historical accounts on the valley's touristic development but they also posed a threat in terms of what they signified of an endless, accelerated and technological development of the valley. They were implicitly involved as such in the *Touristikers*' calculations in the play's last scene. It was probably for them that the cows' strong odour would have to be suppressed, to please them and their more demanding needs for speed and comfort, as Mr Frei repeatedly emphasized.

Hoteliers in particular, whether or not they were involved in the play, readily expressed their scepticism towards group tourism, which they usually framed as economic concerns. The greater amount of cleaning and administrative work necessitated to sustain this type of tourism was a problem. Investments in personnel and time were high, while tourist groups often benefited from discounts, arranged between hotels and tour operators. As a hotelier of the play's organizing team once told me, the hospitality industry market had developed in such a way over the last decades that hotels with over a hundred rooms could no longer survive without opening their doors to the 'masses'. This reality motivated the hoteliers' widespread support for JR's project, along with the hope that it would keep the classical guests in the valley.

Relying on groups was dissatisfying but essential financially in a precarious hospitality industry where the number of regular individual guests had receded because of the difficult financial context and changing holiday patterns.[13] Group tourism was a necessary but short-term investment, given that groups were most likely to never come back to the valley, as I was told repeatedly. This was a tangible consequence of the transformations of the global tourism market towards a faster economy. But as one hotelier – the owner of one of the oldest businesses in the valley – explained, this presence of mass tourism was not just irksome for them and the staff but also for the other, remaining 'regular guests' or *Stammgäste* who wished for a 'familiar, feel-good atmosphere' (*familiäre Wohlfühlatmosphäre*), which was disrupted by the presence of 'groups of fifty people checking in and out'. Like her, other hoteliers (and locals) often shared their concerns that the regular guests would no longer come to the valley because of this new type of tourism. The idea of a symbiosis of tourism as a historical business made up of regular guests now disturbed by 'unfamiliar' guests indeed reflected the opinion of other, individual guests I met in the valley. Bothered by the crowds of tourists, these individuals frequently viewed the new guests as more destructive than themselves.

These accounts point to two phenomenon that are well discussed in scholarly literature on tourism: the tourists' ascription of the guilt of destruction to other travellers (Enzensberger 1996), as well as tourism suppliers' attempts to push for alternative forms of tourism to please dissatisfied tourists and locals (Stronza 2001). Yet, they also reveal what I suggest could be described as a new phase of critique of tourism in which the makers (and consumers) of tourism increasingly evoke growing global discourses on 'overtourism' – a buzzword in the media, policy-discourses and academia since 2018 (Milano 2018; Peeters et al. 2018).[14] Originally formulated by host populations to utter a critique of tourism's detrimental impacts on their communities, these makers of tourism evoked 'overtourism' in order to foster more desirable forms of it. This is also what Yrigoy et al. (2024) have recently discussed in the context of Majorca, where they analyse how hoteliers, tourism planners and other key actors of tourism 'co-opt' such discourse to follow their own (capitalist) interests.

As I spent time in the valley, however, I realized that the negative label 'mass tourism' did not always equate with group tourism, contrary to the definition proposed by many locals. The non-European, non-historical guests from places like Saudi Arabia and South Korea, travellers in small groups and families, and even individual travellers were regularly described as 'mass tourists'. Culturally different, they were associated with problematic customs and behaviour, which made for an unsatisfying, difficult or 'unprofitable' business. Korean honeymooners – an important segment for the valley – were, for instance, often conflated into the anonymized Asian mass tourists because of their features. Similarly, the Saudis, who were in fact some of the most profitable in economic terms, as reports show (HotellerieSuisse 2013), were often described as particularly worrying figures of mass tourism in local informal talk, although they only constituted a small, receding percentage of the region's guests[15] during my first stay in the village. As I will argue in the next chapter, nativist and racializing stereotypes were central in evoking visions of familiarity and local profitability in relation to the Swiss and European guests in the region and the country.

CONCLUSION

Contrary to the image of a passive, individual yearning for a lost past that nostalgia may at first evoke, this chapter has explored the vibrancy of the social project that nostalgia is. Starting from the amateur historical play *Alpine Glow* and tracing the long genealogy and complex history of nostalgia has revealed that the concept of nostalgia has developed in multiple directions. It has described a disease, a feeling, a space, a time, a moral position, an aesthetic, a critique of capitalist modernity, a commodification of this critique, and a tool of social inclusion and exclusion. The developments around nostalgia have spanned centuries and between different contexts but, as I have shown here, it remains an essential concept to understanding life in the Alps, where it found its origins in the seventeenth century. Stemming from the European Romantic elites, this 'outsider' concept has shaped the development of tourism and travellers' expectations of the Alps, but also radically altered the perceptions of inhabitants of the Alps regarding their own environments and identities.

Given its importance and malleability, it is not surprising that a play exploiting the multiple forms and uses of nostalgia emerged directly after the intense discussions and conflicts that took place in the divided valley. In a contentious present, which was the direct consequence of the polarized local opinions on the future of the valley, many of the participants of the play and its audience enjoyed gathering around a common history, probably in search of what Hodges has described as the 'palliative' effects of nostalgia (Hodges 2010). The critical potential of the play, if limited, also united parties that stood in conflict before the summer of 2019: supporters and opponents of the 'project of the future', hoteliers and locals creatively defined what type of tourism was, and was not, desirable. But in a context where tourism made everything, even this endeavour was commodified as a touristic attraction for the European and Swiss tourists, invited to be spectators of this project, as historical consumers of Alpine nostalgia.

The play's afterlife was short. It was a success because the organizers did not draw deficits but represented an achievement for the play's actors, like Thomas, who had been recognized for their skills; nevertheless, it was a punctual event in a context still highly dominated by predatory forms of tourism development. The project's critical evaluation of the present did not directly question the makers of this type of tourism – considered too risky to confront – but instead, focused on its symptoms. The sceptical attitude of locals towards those who embodied a new, accelerated and different form of tourism – non-Western guests or so-called 'mass tourists' – was, however, much more direct, as I will explore in the next chapter.

NOTES

1. *Alpenglühen* was written and directed by Jürg Fankhauser. The name of the association that produced it in German is *Verein Jungfrau Theater*. It was dissolved in 2021. Excerpts from the play are my own translation from German, based on field notes I took while attending it.'
2. See: https://www.rts.ch/emissions/temps-present/2019/video/tourisme-suisse-la-masse-critique-26683759.html (last accessed 14 April 2025).
3. In a media interview, one NGO president played around with the idea of 'mass instead of class' (*Masse statt Klasse*) materialized by the new project, mostly highlighting its aesthetically disrupting impacts: sustainability did not 'look' like JR's project (see Chapter 2). I refrain from citing my sources to align with my anonymization strategy.
4. As of 2010, the European guests represented 44 per cent of the hotel business and 35 per cent of the parahotellerie overnights. In 2019, the year of the play, they represented 23 per cent of the hotel business overnights and 32 per cent of the parahotellerie overnights. In 2010, the Swiss represented 28 per cent of the hotel business and 23 per cent of the holiday homes overnights. In 2011, the Swiss represented 31 per cent of the hotel business and 38 per cent of the parahotellerie overnights. As of 2019, the year of the play, the Swiss represented 26 per cent of the hotel business and 31 per cent of the parahotellerie overnights. In 2010, the Asian market represented 20.5 per cent of the hotel overnights and 2.6 per cent of the parahotellerie overnights. In 2019, they represented 38 per cent of the hotel overnight stays and 17 per cent of the parahotellerie overnight stays. These figures are based on the local tourism office statistics.
5. See Leybold-Johnson 2009.
6. On one of his trips to Switzerland, Mendelssohn wrote the following in his diary: 'Of all the countries I know, it's the most beautiful and the one I'd most like to live in if I

got really old. The people are so happy, they look so happy and so does the country! You are often moved to tears to see that honesty still finds a refuge here and that it has not entirely disappeared from this world; you are happy not to come across any more beggars, gruff employees, in short, to find here the very opposite of what you see elsewhere among men' (Translation by DL, Pidoux 1988).

7. Mendelssohn is said to have written his Symphony No. 11 based on a songbook entitled *Sammlung von Schweizer Kühreihen und Volksliedern* (Collection of Swiss Kühreihen and Folksongs), a book brought back by many travellers after their trips to Switzerland (Gallusser 2021).
8. UNESCO officially recognized the region as global heritage. The 'natural criteria', and 'social and economic activities' such as farming were important parts of the making of the region into a heritage object. See Criteria VII, VIII, IX in https://whc.unesco.org/en/criteria (last accessed 15 April 2025).
9. https://jungfraualetsch.ch/welterbe/welterbe-gebiet/ (last accessed 25 June 2024).
10. Berliner's suggestion that nostalgia is at the core of heritage practices has been discussed by many other authors too (Lowenthal 1989; Di Giovine 2008; Campbell, Smith and Wetherell 2017). On the connections between nostalgia, heritage and tourism, see Frow 1991; Graburn 1995; Dann 1998; Caton and Santos 2007.
11. The other scenes depict further events of the local history of tourism in the 1900s, 1930s and 1970s. I decided to use the first and last scene of the play as they present moments of creation and destruction, which are essential for my argument here.
12. Resembling those of other conservative Alpine regions (see Stacul 2003; Lems 2022, 2023).
13. Hospitality has been considered risky sector, with difficulties in accessing financing and bank loans (Deloitte 2015).
14. Although Sytnik and Stecchi have coined the term 'overtourism' in 2015 (Sytnik and Stecchi 2015). The origins of the term are debated, but its surge in popularity in 2018 has been attributed to its selection as one of the Oxford English Dictionary's words of the year (O'Regan et al. 2022).
15. Between 2017 and 2018, the 27,883 overnight stays of the 'GCC and Africa' guests decreased to 27,009 in hotels (5 per cent of the total stays) and from 32,216 to 30,848 in guesthouses (12 per cent of the total stays). These figures are based on local statistics.

REFERENCES

Angé, Olivia, and David Berliner. 2014. *Anthropology and Nostalgia*. New York: Berghahn Books.

———. 2021. *Ecological Nostalgias: Memory, Affect and Creativity in Times of Ecological Upheavals*. New York: Berghahn Books.

Atia, Nadia, and Jeremy Davies. 2010. *Nostalgia and the Shapes of History*. London: Sage UK.

Azcárate, Matilde Córdoba. 2020. *Stuck with Tourism: Space, Power and Labor in Contemporary Yucatan*. Oakland: University of California Press.

Bachmann, Stefan. 2012. 'Heimatschutz'. *Dictionnaire Historique De La Suisse*.

Berlin, Isaiah. 1976. *Vico and Herder: Two Studies in the History of Ideas*. London: Hogarth Press.

Berliner, David. 2012. 'Multiple Nostalgias: The Fabric of Heritage in Luang Prabang (Lao PDR)', *Journal of the Royal Anthropological Institute* 18(4): 769–86.

———. 2014. 'Are Anthropologists Nostalgist?' in David Berliner and Olivia Angé (eds), *Anthropology and Nostalgia*. New York: Berghahn Books, pp. 17–34.

Betz, Hans-Georg, and Carol Johnson. 2004. 'Against the Current – Stemming the Tide: The Nostalgic Ideology of the Contemporary Radical Populist Right', *Journal of Political Ideologies* 9(3): 311–27. Doi: 10.1080/1356931042000263546.

Bianchi, Raoul. 2018. 'The Political Economy of Tourism Development: A Critical Review', *Annals of Tourism Research* 70: 88–102.

Campbell, Gary, Laurajane Smith and Margaret Wetherell. 2017. 'Nostalgia and Heritage: Potentials, Mobilisations and Effects', *International Journal of Heritage Studies* 23(7): 609–11.

Casey, Edward. 1987. 'The World of Nostalgia', *Man and World* 20(4): 361–84.

Caton, Kellee, and Carla Almeida Santos. 2007. 'Heritage Tourism on Route 66: Deconstructing Nostalgia', *Journal of Travel Research* 45(4): 371–86.

Chambers, Erve. 1997. *Tourism and Culture: An Applied Perspective*. Albany, NY: State University Press of New York.

Cheer, Joseph M., Claudio Milano and Marina Novelli. 2019. 'Tourism and Community Resilience in the Anthropocene: Accentuating Temporal Overtourism', *Journal of Sustainable Tourism* 27(4): 554–72.

Chio, Jenny T. 2014. *A Landscape of Travel: The Work of Tourism in Rural Ethnic China*. Seattle: University of Washington Press.

Colomb, Claire, and Johannes Novy. 2017. *Protest and Resistance in the Tourist City*. Abingdon: Routledge.

Dann, Graham. 1998. 'There's No Business Like Old Business': Tourism, the Nostalgia Industry of the Future', in W. Theobald, *Global Tourism: The New Decade*. Oxford: Butterworth Heinemann, pp. 29–43.

Deloitte. 2015. 'Enquête sur le secteur de l'hôtellerie Suisse'. Zurich: Deloitte.

Denning, Andrew. 2014. *Skiing into Modernity: A Cultural and Environmental History*, Vol. 3. Oakland: University of California Press.

Di Giovine, Michael A. 2008. *The Heritage-scape*. Boulder, CO: Lexington Books.

Duffy, Rosaleen. 2006. 'The Politics of Ecotourism and the Developing World', *Journal of Ecotourism* 5(1–2): 1–6.

Enzensberger, Hans Magnus. 1996. 'A Theory of Tourism', *New German Critique* 68: 117–35.

Franklin, Adrian. 2008. 'The Tourism Ordering: Taking Tourism More Seriously as a Globalising Ordering', *Civilzations. Revue Internationale d'Anthropologie et de Sciences Humaines* 57: 25–39.

Fritzsche, Peter. 2001. 'Specters of History: On Nostalgia, Exile and Modernity', *The American Historical Review* 106(5): 1587–1618.

Frow, John. 1991. 'Tourism and the Semiotics of Nostalgia', *October* 57: 123–51.

Gallusser, Lion. 2021. 'Von allen Ländern, die ich kenne, ist dies das Schönste', *Tonhalle Orchester Zurich*. Retrieved 16 April 2025 from https://www.tonhalle-orchester.ch/News/Mendelssohn-Und-Die-Schweiz-Von-Allen-Laendern-Die-Ich-Kenne-Ist-Dies-Das-Schoenste/.

Geismar, Haidy. 2015. 'Anthropology and Heritage Regimes', *Annual Review of Anthropology* 44: 71–85.

Gibson, Chris. 2010. 'Geographies of Tourism: (Un)Ethical Encounters', *Progress in Human Geography* 34(4): 521–27. Doi: 10.1177/0309132509348688.

Graburn, Nelson. 1995. 'Tourism, Modernity and Nostalgia.' in Akbar S. Ahmed and Cris N. Shore (eds), *The Future of Anthropology: Its Relevance to the Contemporary World*. Atlantic Highlands, NJ: Athlone, pp. 158–78.

Grasseni, Cristina. 2016. *The Heritage Arena: Reinventing Cheese in the Italian Alps*, Vol. 5. New York: Berghahn Books.

Heitmann, Sine, Peter Robinson and Ghislaine Povey. 2011. 'Slow Food, Slow Cities and Slow Tourism', in Peter Robinson, Sine Heitmann and Peter Dieke (eds), *Research Themes for Tourism*. Wallingford UK: CABI, pp. 114–27. DOI: 10.1079/9781845936846.0114.

Hodges, Matt. 2010. 'The Time of the Interval: Historicity, Modernity and Epoch in Rural France', *American Ethnologist* 37(1): 115–31.

Hofer, Johannes. 1934. 'Medical Dissertation on Nostalgia', *Bulletin of the Institute of the History of Medicine* 2(6): 376–91.

HotellerieSuisse. 2013. *L'accueil des hôtes des états Arabes du Golfe*. Lausanne: HotellerieSuisse.

Jameson, Fredric. 1991. *Postmodernism, or, the Cultural Logic of Late Capitalism*. Durham, NC: Duke University Press.

Jankélévitch, Vladimir. 1983. *L'irréversible et la nostalgie*, Vol. 123. Paris: Flammarion.

Leite, Naomi, and Nelson Graburn. 2009. 'Anthropological Interventions in Tourism Studies', in Tazim Jamal and Mike Robinson (eds), *The SAGE Handbook of Tourism Studies*. London: Sage.

Leitenberg, Danaé. 2022. 'Romantic Visions of the Village: Tourism, Nationalism, Locality and the Politics of (Non-)belonging in the Swiss Alps', *Quaderns de l'Institut Català d'Antropologia* 38(2): 195–212.

Lems, Annika. 2016. 'Ambiguous Longings: Nostalgia as the Interplay among Self, Time and World', *Critique of Anthropology* 36(4): 419–38.

———. 2022. 'Deciphering Everyday Meaning-Making with Gramsci', *Dialectical Anthropology* 46: 395–415. Doi: 10.1007/S10624-022-09658-5.

———. 2023. 'Anti-Mobile Placemaking in a Mobile World: Rethinking the Entanglements of Place, Im/Mobility and Belonging', *Mobilities* 18(4): 620–34. Doi: 10.1080/17450101.2023.2220942.

Leybold-Johnson, Isobel. 2009. 'Mendelssohn Composes a Swiss Legacy', *Swissinfo*. Retrieved 15 April 2025 from https://www.swissinfo.ch/eng/culture/mendelssohn-composes-a-swiss-legacy/998460#:~:text=Less%20well%20known%20perhaps%20are,dance%20tune%20from%20the%20Emmental.

Lowenthal, David. 1989. 'Nostalgia Tells It Like It Wasn't', in Christopher Shaw and Malcom Chase (eds), *The Imagined Past: History and Nostalgia*. Manchester: Manchester University Press, pp. 18–32.

MacCannell, Dean. 2013. *The Tourist: A New Theory of the Leisure Class*. Berkeley, University of California Press.

Mathieu, Jon, and Simona Boscani Leoni. 2005. *Die Alpen! Zur Europäischen Wahrnehmungsgeschichte Seit der Renaissance = Les Alpes! Pour une Histoire de la Perception Européenne depuis La Renaissance*, 1. Aufl. Ed. Vol. 2, *Studies on Alpine History*. Bern: P. Lang.

Milano, Claudio. 2018. 'Overtourism, Social Unrest and Tourismphobia: A Controversial Debate', *PASOS Revista de Tourismo y Patrimonio Cultural* 16(3): 551–64.

Milano, Claudio, Marina Novelli and Joseph M. Cheer. 2019. 'Overtourism and Tourismphobia: A Journey through Four Decades of Tourism Development, Planning and Local Concerns', *Tourism Planning & Development* 16(4): 353–57. Doi: 10.1080/21568316.2019.1599604.

Mostafanezhad, Mary. 2014. 'Volunteer Tourism and the Popular Humanitarian Gaze', *Geoforum* 54: 111–18.

Neveling, Patrick, and Carsten Wergin. 2009. 'Projects of Scale-Making: New Perspectives for the Anthropology of Tourism', *Etnográfica. Revista do Centro em Rede de Investigação em Antropologia* 13(2): 315–42.

Nöbauer, Herta. 2022. 'Our Existence is Literally Melting Away', in Susanna M. Hoffman, Eriksen Thomas Hylland and Mendes Paulo (eds), *Cooling Down: Local Responses to Global Climate Change*. New York: Berghahn Books, pp. 223–44.

O'Regan, Michael, et al. 2022. 'Unpacking Overtourism as a Discursive Formation through Interdiscursivity', *Tourism Review* 77(1): 54–71.

Pasieka, Agnieszka. 2022. 'Theft of Gramsci? On the Radical Right, Radical Left and Common Sense', *Dialectical Anthropology* 46(4): 417–36. Doi: 10.1007/S10624-022-09681-6.

Peeters, Paul, et al. 2018. 'Research for TRAN Committee – Overtourism: Impact and Possible Policy Responses'. A European Union Study.

Petite, Mathieu. 2011. *Identités en chantiers dans les Alpes: Des projets qui mobilisent objets, territoires et réseaux*, Vol. 20. Bern: Peter Lang.

———. 2013. 'Les Montagnards face aux écolos. Une énième déclinaison au travers de l'initiative Weber', *Journal of Alpine Research* (Hors-Série).

Pidoux, Edmond. 1988. 'Le Regard de Mendelssohn', *Club Alpin Suisse*. Retrieved 25 February 2025 from https://www.sac-cas.ch/fr/les-alpes/le-regard-de-mendelssohn-12548/.

Robertson, Roland. 1992. *Globalization: Social Theory and Global Culture*, Vol. 16. London: Sage.

Rosa, Hartmut. 2013. *Social Acceleration: A New Theory of Modernity*. New York: Columbia University Press.

Salmose, Niklas. 2019. 'Nostalgia Makes Us All Tick: A Special Issue on Contemporary Nostalgia', *Humanities* 8(3): 144.

Schär, Bernhard. 2012. 'Bauern und Hirten Reconsidered. Umrisse der "erfundenen" Schweiz im imperialen Raum', in Patricia Purtschert, Barbara Lüthi and Francesca Falk (eds), *Postkoloniale Schweiz. Formen Und Folgen Eines Kolonialismus Ohne Kolonien*. Bielefeld: transcript, pp. 315–31.

———. 2015. 'On the Tropical Origins of the Alps', in Patricia Purtschert (ed.), *Colonial Switzerland: Rethinking Colonialism from the Margins*. London: Palgrave Macmillan, pp. 29–49.

Seiler, Roman. 2016. 'Die Jungfrau-Region boomt – Rund 70 Prozent der Touristen stammen aus Asien', *Aargauer Zeitung*. Retrieved 10 May 2020 from https://www.aargauerzeitung.ch/Schweiz/Die-Jungfrau-Region-Boomt-Rund-70-Prozent-Der-Touristen-Stammen-Aus-Asien-130446719.

Sidali, Katia Laura, and María De Obeso. 2017. 'Successful Integration of Slow and Sustainable Tourism: A Case Study of Food Tourism in the Alpine Region of

Algovia, Germany', in Michael Clancy, *Slow Tourism, Food and Cities*. New York: Routledge, pp. 169–80.

Stacul, Jaro. 2003. *The Bounded Field: Localism and Local Identity in an Italian Alpine Valley*. Vol. 18. New York: Berghahn Books.

Starobinski, Jean. 1966. 'The Idea of Nostalgia', *Diogenes* 14(54): 81–103.

Stewart, Kathleen. 1988. 'Nostalgia – a Polemic', *Cultural Anthropology* 3(3): 227–41.

Stone, Alison. 2014. 'Alienation from Nature and Early German Romanticism', *Ethical Theory and Moral Practice* 17(1): 41–54.

Stronza, Amanda. 2001. 'Anthropology of Tourism: Forging New Ground for Ecotourism and Other Alternatives', *Annual Review of Anthropology* 30(1): 261–83.

———. 2005. 'Hosts and Hosts: The Anthropological of Community-Based Ecotourism in the Peruvian Amazon', *NAPA Bulletin* 23(1): 170–90.

Summermatter, Stephanie. 2010. 'Protection de la nature', *Dictionnaire Historique de la Suisse*.

Sytnik, Oxana, and Francesco Stecchi. 2015. 'Disappearing Coastal Dunes: Tourism Development and Future Challenges. A Case Study from Ravenna, Italy', *Journal of Coastal Conservation* 19: 715–27.

Tucker, Hazel. 2005. *Living with Tourism: Negotiating Identities in a Turkish Village*. London: Routledge.

Urry, John, and Jonas Larsen. 2011. *The Tourist Gaze 3.0*. London: Sage.

Wodak, Ruth. 2015. 'The Politics of Fear: What Right-Wing Populist Discourses Mean'. London: Sage.

Yrigoy, Ismael, et al. 2024. 'Co-Opting Overtourism: Tourism Stakeholders' Use of the Perceptions of Overtourism in Their Power Struggles', *Journal of Sustainable Tourism* 32(4): 818–34. Doi: 10.1080/09669582.2023.2178445.

Zimmer, Oliver. 1998. 'In Search of Natural Identity: Alpine Landscape and the Reconstruction of the Swiss Nation', *Comparative Studies in Society and History* 40(4): 637–65.

CHAPTER

4

NAVIGATING TOURISM DEPENDENCY

'To live where others spend their holiday', says Franziska, admiring the breathtaking panorama of the Swiss Alps before telling me what a luxury she thinks this is, sometime in winter 2018. Now in her mid-sixties, Franziska describes herself as a 'native local' according to the local politics of belonging that I defined at the outset of this book. She has a local family name, speaks local dialect, was born in the valley and has been living here her whole life. Franziska and I are sipping our coffees on the terrace of a winter ski bar between two ski descents on one of her days off, before she has to go home to prepare lunch for her husband, a JR employee like her three adult sons, and then prepare the holiday rental for the guests who are about to arrive. This is what she does on the three days a week she does not have to work as an accountant for one of the ski rental companies of the valley. The guests who will arrive in a couple of hours are *Stammgäste* or regular guests, she tells me, estimating that this couple from Germany have been coming to the valley for approximately twenty years. It is still early and we are almost alone, a privilege for Franziska. The crowd of skiers and tourists is about to arrive, but, for now, I also feel privileged.

Later on that day, Franziska and I are chatting in her kitchen – her 'little paradise', as she calls it. Images of farming times and values she cherishes decorate the walls of her home, in a house inherited by her husband, who hails from 'an even more native family' than hers. She tells me something she will repeat often over the next few years of my research: she genuinely wishes she had lived in the past, in a previous era she has never even witnessed, before tourism and the 'fast, non-committal, hurried times' (*schnelle, unverbindliche, hetzige Zeit*), as she puts it, 'anyone can just get in their car and leave at any moment'. 'We are anonymous to each other.' In pre-tourism times, her ancestors were poor, which was synonymous with greater solidarity and social cohesion. 'People are doing too well now', she muses. Tourism has improved the villagers' financial situations but also eliminated the need for them to come together. 'Tourism and the constant coming and going of tourists has probably made locals *unruhig* (restless)', she says.

Franziska's nostalgic evocations and her dislike for the present echoed Thomas's and Moritz's critiques, as well as those formulated in the play. But Franziska interspersed such rosy pictures of the Alpine past and her critique of modernity with statements arguing for the absolute need to build new infrastructure to reach the peaks faster and secure the leading position of her village. 'We have overslept', she told me regularly, echoing Mr Frei's words. Franziska's nostalgic dreams faded away as she told me that the 'capital' of the beautiful region needed to be captured and innovated further in order to attract more guests. The new cable car project would foster the common good, even if Franziska admitted not needing it personally. 'I am scared by the dimensions of the project', she said, 'but it is needed'. In our discussions, Franziska often referred to her three sons and her husband as JR employees and enthusiastic supporters, and echoed their convictions. 'It is not important for me personally, but for the youth here . . . what would they do without the cable car?' she asked me, equating JR development with tourism development. If tourism were to disappear, a possibility Franziska reasoned with regularly, 'I would have to get cows, goats and chickens', she said once, hinting at her cat sleeping next to us. She enjoyed the company of animals that were pets, but would have to return to farming and cattle without JR and tourism.

Drawing on these two imaginaries of preservation and innovation, Franziska pointed to the landscape around us on the summer's day when I came to interview her. The *Dreigestirn* mountains before our eyes, she stated, were an extraordinary resource that needed to be exploited to continue offering comfort and experiences to the guests. By contrast, the mountain next to it, the valley's *Hausberg* or home mountain, just above Franziska's house, marked the beginning of what she described as 'a green belt', which should remain 'untouched'. While Franziska was a proud native villager, she defined places that were less village-like than others within the valley, places that were 'safe to innovate' because they had already been corrupted, already dedicated to development – the *Dreigestirn*, the side of the valley where Beat lived, the valley floor below us, the village centre around the *Hauptstrasse*. Franziska's garden, its colourful flowers and lush green plants, were undoubtedly part of the purer, timeless side of the village, a sentiment shared by guests: 'The tourists are taken aback when they arrive here', Franziska told me proudly. After a drive up the valley's flanks and into the scattered settlement (*Streusiedlung*), her guests entered the authentic Alpine experience upon arrival at her home. Franziska believed that, by coming to the Alps, they broke away from the lowlander life she imagined for them – one made of computer screens and stressful times. In doing so, she mobilized the dominant escapist tourism imaginaries of the region, inherited from the Romantics. However, Franziska explained that on some days the peacefulness of her place was undermined: depending on the direction of the wind, she could hear the screams of tourists, mostly from Arab countries, going down the mountain on scooters or using the summer sledge on one of the other valley's flanks. After working in the garden or in her house, she sometimes stopped to count the increasing number of black vans – typically rented out by Arab guests – that came to her side of the valley.

Listening to Franziska, I often felt lost. Her romanticized evocations of the Alps were often abruptly followed by marketing slogans, which left me confused. This spoke of another, more visual, sort of puzzlement I experienced when witnessing the juxtaposition of the latest state-of-the-art infrastructure with nostalgic imaginaries of Alpine authenticity (Leitenberg 2018). Franziska, like the valley, seemed to embody the contradictory 'chronopolitics of tourism' (Norum and Mostafanezhad 2016), consisting of both the capitalist future-making and the romanticizing views of the past that I have explored up to this point. Tourism dependents like her ambiguously participated in an extractive capitalist industry motivated by necessity, all the while often expressing strong doubts about it, notably informed by the nostalgic narratives of tourism. Both developmental and conservative, Franziska's position as a tourism dependent escaped clear definitions of 'opponent' or 'supporter' of the growth and acceleration promoted around the cable car. She embodied a middle-ground position that, as puzzling as it was, became impossible to discard for me as fieldwork progressed. Her experience not only echoed many discussions I had with locals,[1] but it also illustrated a deeper reality: it was paradigmatic of a tension between two, oppositional temporal demands of tourism – between a race towards the future and an effort to preserve the past.

Paying attention to the everyday, lived aspects of tourism dependency, in this chapter I reflect on the experience of Franziska, especially with respect to her capacity to attend to and to navigate the ambiguous, uncertain character of tourism. Drawing lines between times, spaces and guests that are either deemed positive, meaningful and satisfying, or negative, alienating and upsetting, she demonstrates that 'the toured ones' are far from passive subjects. However, between new developments pertaining to globalized, capitalist tourism and the long-standing politics of nativism, this chapter also shows how Franziska's navigation seemed increasingly difficult to undertake.

Using the word 'navigation' here, I rely on anthropologist Henrik Vigh's work, which has addressed how actors live through environments or situations of 'social volatility and opacity' (Vigh 2009: 419). For Vigh, navigation is 'motion within motion', referring to 'how people move and manage within situations of social flux and change' (420). Volatility and opacity particularly mark tourism dependency, where the drivers, transformations and consequences of tourism appear to be difficult to 'control' for its makers or suppliers, as I have previously shown (chapters 2 and 3). In this chapter I show how this also holds true for locals like Franziska, Thomas and Moritz who, while less involved in the planning, shaping or execution of tourism development, encounter tourism as a challenging reality – one that is difficult to comprehend and live with or adapt to and thus requires skilful 'navigation'.

COPING WITH TOURISM

One day, after Franziska and I had come back from a short trip to the *Hauptstrasse* to run errands, she turned to me with a hand on her chest and a frown on her face and said, 'See what going to the village centre does to me'. The traffic noise, the shouting, the rhythmic tapping of tourists' ski boots on the pavement, the overcrowded sidewalk alongside the steady stream of affluent traffic and the pungent smells of petrol mixed with odours wafting from the air vents of hotels and restaurants made her literally feel sick – it made her head ache and twisted her stomach. She regularly told me she believed that the valley was the most beautiful place in the world, but it seemed that not *all* the valley was worthy of this description.

In contrast with the caring vocabulary Franziska used to describe 'her' part of the valley, the centre – located approximately forty minutes away by foot, below her house – was the place of malaise, dissatisfaction, absurdity even. The centre recalled the anonymity and crowds that she thought were so typical of the lowlands: bus drivers, waiting staff and salesmen even mistook her for a tourist, which was a great offence for the native that Franziska was. Coming back home, she was able to breathe and enjoy what she described as her luck: 'If you have a green belt around you, if you have barn, are able to spend time with animals and let off steam in the garden, going to the mountains when you want and how you want – I find that is an extreme privilege'. Those in the centre – tourists but also locals, mostly outsiders in Franziska's descriptions, who lived closer to or in the crowds and concrete – did not enjoy the same privilege. She often implied that they were leading alienated lives shaped by technological progress but existential vacuity. Franziska, by contrast, was proud to call herself a 'hillbilly' because this meant being connected to the 'real' Alpine farming values.

Franziska confirmed what Anna and many others had described to me as a resource of the *Einheimische*. Although Anna was bothered by the growing phenomenon of mass tourism, she told me confidently that 'there will always be places for the *Einheimische*'. Tourists gathered in so-called hotspots, leaving the rest of the valley untouched and to be enjoyed by locals who knew where to go, Anna contended. The play *Alpine Glow* relied on this tacit knowledge too: it was set up off the beaten tracks and outside of the busiest times of the day. It was accessible to Swiss and European tourists provided they had been given the correct information at their hotels, but not to 'the masses' or 'the Asians', who stayed around the valley's main street during the day. Thomas, too, drew boundaries between the authentic, natural and picturesque village and the corrupt, upsetting, urban village, associated with the lowlands and the city. While Franziska felt literally sick after trips to the village centre, Thomas experienced fear (*Angst*) in it, seeing the crowds and the countless construction sites, he said hitting his chest multiple times when we met on the *Hauptstrasse*.

I often tried to picture the precise boundaries drawn by these locals between an authentic and an overtouristified valley. Like Franziska, Moritz and Beat, a majority of the native locals I met lived high up on the valley's sides, often more than 1,000 metres above sea level, in comparison to the 900 metres of the train station of the lower valley floor (below 900 metres). Their houses were located on the outer edges of the scattered settlement (*Streusiedlung*), before these turned into a dense and elongated centre around the main street or *Hauptstrasse*. However, except for the one-kilometre-long portion of the village's main road that was unanimously decried as a place of excess, chaos and noise, and more recently the new cable car terminal in the valley floor, resembling an airport, the border between what was considered genuine or natural and what was artificial and new was difficult to locate. Thomas's house, for instance, which he believed was located in a place of authenticity and integrity, would already be considered 'too much' for someone like Franziska because of its greater proximity to the centre.

What such shifting, subjective boundaries nonetheless reveal is the persistence of a practice of defining one's property as the authentic centre of the world, especially in comparison to an 'outside' marked by destruction. The Swiss urban lowlands and their tall concrete buildings, were places where 'corpses could be laying for days unnoticed' because of the lack of 'solidarity', as native Andreas for instance told me, echoing with Franziska's feelings towards what she imagined of Berne or Zurich. But this also affected the centre of the village, its traffic, crowds and malls, all examples of the disturbing *Verbetonierung* or excessive concretization of the Alps – an image present in the village, illustrated in the theatre play for instance, but also beyond (Petite 2013).[2] Instead of agreeing on the same spatial division of the valley, many native locals shared a perspective on the valley: that of ordering the world between one's safe, harmonious home and the threatening, strange, unhomely outside.[3] Franziska's account of the 'perfect world' she inhabited points to a form of coherent whole that aligns with the notion of the *Heimat* that I briefly presented in Chapter 1.

From its origins, the concept of *Heimat* was linked to peasantry, farming and the ideal of an autonomous individual world (Cole and Wolf 1999; Leimgruber 2019). Like Alps, the *Heimat* has often been the site of Romantic longings, a place of authenticity in an alienating world of disturbance and dissatisfaction (Jaeggi 2014: 9). Translated to Franziska's case, the green belt, the historical part of the village and the valley's *Hausberg* (home mountain) were all made into safe and pure prolongations of the *Heimat*, while the cable car, malls and the terminal were marked by external, urban influence. This echoed the thoughts of Christine, another native local and a relative of Franziska, who exclaimed as we overlooked the village, 'We are not a town (*Stadt*) and we do not want to become one!' She was visibly upset about what I had just told her: an 'outsider' (*Auswärtiger*) I talked to who worked in one of the village's sports shops had described her native village as a town, as he hinted at the number of people on the streets and the global connections that marked the place. Although this depiction bothered her, Christine also conceded that the overcrowded, congested centre indeed felt different from what a rural village should feel like – and was the reason why she herself avoided this part of the valley.

While the *Heimat* features a particularly important role in these narratives, the local preoccupations they express can also be found among inhabitants of other, non-German-speaking Alpine areas, as the work of anthropologist Jaro Stacul in the Italian Vanoi Valley has prominently revealed (Stacul 2003a, b; 2006). Stacul's interlocutors articulated a pronounced sense of division between the valley – shaped by its history under Austrian rule and its relative economic underdevelopment – and the broader Italian context, including neighbouring villages. Simultaneously, Stacul examines how internal divisions, particularly those surrounding land ownership and private property, significantly influenced his interlocutors' primary concerns. His concept of the 'bounded field' encapsulates not only individual commitments, practices and values but also the broader social and political relationships structured around ideals of autonomy, order and control (Stacul 2003a).

Both my Swiss field site and the Vanoi Valley highlight the intricate dynamics of Alpine villages, shaped by their internal spatial boundaries. Such boundaries informed many of my native interlocutors' movements through the valley: necessary errands were usually run outside of the busiest times of the week or the day in a region where shops had long opening hours to welcome tourists. Some native locals explained that they also avoided the main village street if they could when they had to reach the other side of the valley, taking a less busy but longer alternative. When they went to spots that they knew were popular among tourists, they did so very early in the morning, before the crowds – like Franziska described in the introduction of this chapter, and Thomas in the previous one. This reality of avoidance was reinforced by the disappearance of places 'for the *Einheimische*' in the centre, meaning places run by locals with a *Stammtisch* (a regulars' table). 'It's no longer for us', one local told me as we were talking about the centre and its recent developments: fancy hotels, restaurants and sports shops were not places where native locals could be themselves and enjoy the *Feierabend* (after-work time) with peers.[4] Between the consequences of gentrification and their unease towards the centre of the valley, most natives I met thus retreated into their homes up on the sides of the valley (see also Chapter 1).

These accounts reveal the symbolic but also embodied or corporeal 'navigation' (Vigh 2009) of tourism's ambiguities. While Anna, Franziska and many others knew that much of what they cherished and held dear in in the village – including the possibility to stay in it – was 'cross-financed' by the overcrowded, overbuilt centre and the more intensive forms of tourism it stood for, they navigated this reality by adapting their movements in it. This echoes the practices of 'coping with' or 'resisting' tourism discussed by anthropologist Jeremy Boissevain three decades ago in the context of European seaside destinations and mass tourism (Boissevain 1996; Boissevain and Selwyn 2004). These discussions have, as indicated in the previous chapter, recently re-emerged following the global waves of protest against overtourism (Gössling, McCabe and Chen 2020). The hosts, as Boissevain and the many authors of tourism contestation have shown, are not passive tourism dependents but attempt to live a life off tourism that is the most viable. They do so by 'develop[ing] strategies to protect themselves from tourists bent on penetrating their back regions to stare, undergo authentic experiences and photograph' (Boissevain 1996: 14). They use visible means such as 'hiding, fending, ritual, organized protest and aggression', or what Boissevain names 'covert resistance', meaning practices of 'sulking, grumbling, obstruction, gossip, ridicule and surreptitious insults', which 'avoid direct defiance' (ibid.). According to him, this strategy is a manifestation of the ambiguous position of tourism dependents, living off tourism and the 'goodwill of foreign visitors', while also struggling with the industry's impacts on their daily lives. In this context, covert resistance allows them to express critique without 'confronting [tourists] directly' (ibid.). In the Bernese village, putting on the play *Alpine Glow* can be seen from the perspective of covert resistance, as a response to a situation of dependency on (so-called 'mass') tourism that was experienced as problematic. The play explicitly avoided being confrontational because of the proximity of the organizers and the *Touristikers*, but it did trace a boundary between creative and destructive forms of tourism.

Subscribing to this differentiation between historical and new forms of tourism, locals like Franziska drew not only a temporal but also a spatial line between places that were preserved and connected to a greater past and those dedicated to innovation. Franziska described the conditions that allowed for the strategy of division and avoidance to happen: she enjoyed the 'privilege' of living 'rooted in her *Heimat*' in the symbolic sense of the word but also in its more material terms. She and her husband were the owners of their property – a home they had shaped to their desires and that served as an anchor for their movements in the valley. All of my native interlocutors reported that they enjoyed similar conditions, as they had inherited houses or had access to property via local networks. In a context where property was costly, to be able to retreat in one's own place was even more of a privilege. Although locals like Franziska did not own the means of production of tourism and they were largely unable to control the way tourism was made and shaped in the village, their properties allowed them to separate the valley into past-like, preserved and homely zones and alienating areas sacrificed for the future, which they would avoid.

A GROWING SENSE OF DISPOSSESSION

Tourism growth is a contested process and has been for a long time. In Chapter 1, I showed how, in spite of the progress it promised, touristification in the form of infrastructural sprawl has been a contentious topic in the valley from its origins as a touristic destination, when locals felt thrown into an industry that was upsetting their lives and environment. In a landscape of great unevenness between the local centre, with its hotels and rich tourists, and the outskirts, marked by farming and poverty until the 1950s, there are reasons to believe that many locals have long avoided parts of the valley that were seen as the most touristic, upsetting and 'not for them'. In a time of greater financial precarity and lesser capacity to move through the valley and the world, this avoidance was surely not entirely of their making. This was different from the conscious strategies adopted by locals during my fieldwork.

However, during the course of my research, it seemed that the practical navigation of tourism's daily struggles via divide, avoidance and retreat had lost some of its efficacy for its dependents. The boundary between critical opponents and enthusiastic supporters became increasingly difficult for me to define, as locals I met had a feeling of being overwhelmed following the construction project 'of the century'. In spite of years trying to assess the cable car's effects, many seemed to have been caught off guard, surprised and disappointed by the project. Thomas changed his mind and claimed many of his peers shared his concerns. Franziska also expressed a growing discomfort regarding the new infrastructure. During the construction, she confessed to me that she hoped for tranquillity after villagers had gone through years of contestation and appeased tourism's hunger for innovation. Accepting the cable car was a sacrifice that, if necessary, deserved to be acknowledged and treated with gratitude and sensitivity from the *Touristikers'* side. But Franziska's wishes for stability and peacefulness clashed with a harsh reality. The 'project of the century' had crossed new frontiers of extraction in unexpected and tangible ways.

Starting with the giant hole that had been dug on the valley floor for the construction of the new project's terminal, with three hundred workers involved in the project over nine hundred days. Large convoys of material imported in the valley regularly blocked its main, narrow and winding road, already reaching the limit of its capacity before the project. Many of the locals I met said they had not anticipated the level of disruption to the valley's life that the construction would bring. During one of my visits to Franziska's, one of her nephews – albeit a fervent supporter and worker of JR – shared his impression of a situation involving a friend. He recounted that his friend felt pressured to accept unforeseen modifications to her property following a change of plans regarding the new cable car. If she did not accept, the friend would risk expropriation, according to Franziska's nephew. Franziska's husband, faced with the concerns raised by his nephew and wife regarding his employer appeared uncertain about how to respond and struggled to justify JR's perspective. He ultimately and abruptly steered the conversation away, stating that the project now had to move forward for the common good, and that it had to be done fast. It was not the time to dwell on what he regarded as details and differing interpretations of events.

The story of another local I met recalled this event. Initially in favour of the project for his children – 'they are great JR fans' – Tobias retrospectively felt that he had misjudged the project, and what it meant for his activities and the valley. As one of the landowners impacted by the project, he was presented with new conditions set during the construction, requiring additional modifications and investments from his side. Although Tobias had agreed to his property being partly demolished, the details regarding how, and the extent to which this had to happen, had shocked him. 'Even as I thought that we had an agreement and things were set, apparently they were not. I should have insisted more', he said. Tobias felt pressured into accepting these conditions in order to complete the cable car on time. He concluded that maybe he was not so much of a supporter after all because of the overwhelming scale of the project and its dimensions.

The potential opened up by the 'project of the century' also attracted economic players to the valley, whose intentions to derive profit from the new cable car were met with great scepticism by locals. Shortly after the beginning of the construction, in August 2018, a Swiss construction group presented a new hotel project that was to be built next to the cable car terminal during an information gathering I attended. The hotel project was designed to speak to 'multicultural guests', aiming to offer a holiday experience to young and group tourists with a fully digitalized offer. The hotel would have its own booking platform, digital check-in accessible via smartphone, and a tablet from which one could control features of the rooms and book outdoor activities from the comfort of one's room. Human presence was to be limited to a minimum to respond to the digital habits of young and non-Western tourists. JR supported this business model. Several local hoteliers, feeling that their activities might be impacted by this new player, voiced strong reservations and critiques during the gathering. To locals like Franziska, this proposition of the hotel embodied some of the worst trends of the twenty-first century, along with the anonymization of relations. She, like Thomas, also struggled to understand the relation between the hotel ruins in the centre of the village and the new construction projects. Franziska, like many, regretted that the abandoned buildings were not being given priority. Renovating, she explained, was more costly than building anew and she knew that locals like her did not have a say on what happened in the valley because these were private properties, even if it was their village. Even the municipality admitted having little power over the renovation of hotel ruins which impacted on the image of the village.

Many locals I met who had despised Beat's position before (and sometimes after) the cable car's construction expressed increasingly similar concerns of feeling dispossessed by developments impacting their village, their properties and their activities. Franziska, who was quick to describe Beat as a backward mountain dweller and an enemy of tourism, formulated sentences that echoed with his, wondering whether the locals *really* profited from tourism in the way it had developed, be it because of the crowds or because of the impact of the infrastructural sprawl on her peers. The feeling of disappointment, if not betrayal, that Beat experienced seemed to be increasingly shared, even among those like Franziska and her family, who had previously suggested they, as tourism dependents, should accept making a sacrifice for the common good. While the construction was located in the area of the valley that was designated as 'safe to innovate', it still seemed to affect Franziska, even within the confines of her own 'paradise'. One winter day in 2022, Franziska also told me in disbelief that she had heard that tourism lobbyists were pushing for the construction of more hotels and the coming of more guests. 'Even more tourists? Where will we put them?', she asked me, echoing the mayor's complaints in *Alpine Glow* (Chapter 2). In this context, Franziska was forced to see that the locals' sacrifices had gone largely unrecognized; the race remained relentless as she heard *Touristikers* continually argue about absolute needs for growth, even after the construction of the project. After years of standing still and drama, things were finally going according to plan with the construction of the cable car. But from what I heard from the locals I met, even supporters, the plan itself had become distressing to many. Villagers like Franziska, who wished to foster security via innovation in an unstable business, now increasingly identified thriving capitalist forces as disorientating, disruptive or threatening.[5]

NEW INTRUDERS

Over the years of my research, Franziska explained that her paradise was being confronted with new problems. If she carefully expressed criticism towards the *Touristikers* and although she was indeed shocked to see the changes occurring in places of the valley that she believed were 'safe to innovate', she expressed an increasing sense of being intruded. In the Alpine huts where she would usually enjoy an atmosphere of greater equality and camaraderie between hikers and locals, who would regularly address each other without polite niceties – 'perhaps because we are higher up, closer to God', she said – she told me she had been recently confronted with a new phenomenon. 'Asians' showed up in an area that was previously only visited by the Swiss and the Europeans, 'asking for water for free'. 'What were they doing there?', she asked me, irritated.

Even Franziska's home had now become the object of intrusions: Google indicated that her house was a hotel and had marked one of the small roads adjacent to her house – although barred to non-residents – as accessible to traffic. Tourists she was usually careful not to bump into or host – by, for instance, not putting her holiday rental on the website Booking.com – were suddenly standing in front of her door, or worse, *inside* her house. One day, she found an Arab man standing on the staircase, asking for a hotel room. She had to reprimand him and told him that this was 'private property', Franziska said, complaining about Google and the guest's attitude, his lack of respect for property and boundaries. She therefore asked if I could contact Google to request them to change their GPS data, as she had already tried in vain. Following this incident and a robbery, Franziska now locked her door more regularly, something she never did before – a custom in the village, she said – but perhaps also a sign of 'mountain-dweller naivety'.

Franziska's unfortunate experience of inaccurate GPS data was not an isolated problem; it impacted the whole valley. Tourists drove up as far as the peaks on roads reserved for special buses and residents who had a special permission because their navigation systems indicated they could do so. Aware of the problem, but unable to alter GPS data, the municipality opted for a solution I had never seen before in Switzerland. On many of the valley's streets in 2018, red and yellow signs appeared that declared 'Driving prohibited! GPS data incorrect! No driveway for cars!' in English, German, Chinese, Korean and Arabic languages, with pictograms of cars and motorcycles crossed through.

Other locals complained of instances of intrusion that were not related to GPS indications. One local couple I met was trying to assess the potential trespassing of guests by looking out of their windows onto their garden. After we had finished the interview, the man said to his wife, 'Ah see, now they are leaving', referring to guests who were now walking across the grass surrounding their property to go back to the train station in the valley floor, from which they had wandered to take a photo. The couple then told me that this was not the first time this had happened: they had in fact witnessed more of these events in recent years. They once came across Indian tourists resting on their terrace one evening, and the same thing happened to their daughter, who lived on another side of the valley. On another occasion, they found Asian guests peeking into their living room. The local woman turned to me and concluded with the famous saying: 'There are the travellers (*Reisende*) and the ones being travelled (*Bereiste*). We are the toured ones'.

Complaints about the way new guests especially moved around the valley, in private and public areas, were widespread. The traffic of rental cars – often black vans, recognizable because of their number plates from the cantons of Appenzell (AI) or Vaud (VD), or the cities of Munich (M) or Vienna (W) – was the topic of many complaints because of the 'annoying', 'reckless' driving style of their drivers. These were mainly Arab, Indian or Asian guests, sometimes also (mistakenly) referred to as 'mass tourists', as mentioned in the previous chapter. Regional newspaper articles for instance covered the 'problem' of dangerous Arab tourist drivers, who were described as either ignorant or disrespectful.[6] Images of rental cars, stuck in staircases at the train station or on railway tracks, elicited reactions from Swiss locals, who both ridiculed and despised these behaviours, described as consequences of a new era of intensive tourism and a 'vanishing low season'.[7] More recently, the valley's tourism office responded to the increasingly discussed and mediated issue of overtourism by starting an awareness campaign in German, English and Arabic with instructions for tourists regarding driving, traffic signs and parking but also privacy, highlighting that it was essential not to enter private property, even for a quick photograph.

Moritz described how the AI drivers, 'mostly Arabs', would not respect any rules because they drove up the valley's sides 'with their BMW, Mercedes or Audi' to reach the hotspots, blindly following the GPS instructions. 'When they go past my house, I whistle and say, "Where are you going?" [in English]. They then say they want to go up, and I say, "Turn around, this is a private road". They usually tell me they have never come across such signs and I then respond, "You haven't seen these two signs? The signs look the same across the whole world!"'. Moritz continued: 'They don't give a shit – they do what they want. They are guests, yet they do what they want'. When Moritz complained at the municipality, they told him they could not do anything more than put up the signs. Fencing off whole areas, especially on the sides of the valley, was not possible because doing so would infuriate the farmers. Moritz agreed with this reasoning but then suggested they 'contact Google', to which the municipality retorted that they could not do anything. 'Then I told them to ban it at cantonal level. Someone has to tell [the tourists] that there is a driving ban, or call the car companies and tell them. They don't give a shit about our laws.'

When I met Moritz after the Corona pandemic, he felt, like many,[8] that the intrusions had intensified: he had come across strangers three times in areas that were extremely isolated. 'They didn't even know where they were and were looking at their phones', as they spotted a path that Moritz believed was only accessible to the genuine, native locals like himself. 'They were in nature where no one else is, they went through the forest. There used to be lots of mushrooms there and now there are only a few. In places that are very quiet, people suddenly show up with mobile phones. In some spots there are AI-cars every fifty metres. It used to be quiet . . ., these places used to be beautiful.' Moritz's complaints about AI cars and those who used them dominated most of my exchanges with locals in the valley, especially after the pandemic. Native locals had come up with jokes about them, saying that the letters AI stood for *Alles Iheimisch* ('all native' in local dialect), or *Aus Interlaken* (from Interlaken). As an *Einheimische* villager once told me, these were all jokes, playing on irony: these guests, more than any other, were the opposite of natives.

Against such a feeling of increasing intrusion by mostly 'new' tourists, locals like Franziska attempted to fence off or hide their properties – by not making them visible on booking websites, by putting up signs, by locking external doors, or by telling people off. Yet, she still opened her door to European tourists, mostly German and Swiss, which she estimated to be, respectively, 90 per cent and 10 per cent of her clientele. Franziska had one bad experience with a couple from another country – she vaguely remembered them being from Asia or the Arabic peninsula – who flooded her bathroom: 'They just don't know how to use our facilities', she said, shrugging her shoulders as if to signify an unsurmountable difference. Franziska had had to clean the whole day after their departure, and the money she received in exchange was just not worth her efforts.

These guests also had little in common with her, not just culturally but also as individuals: walking around with their phones, buying ready-made meals and doing everything quickly was an attitude that did not resonate with her, or many of her peers – the alienated and alienating crowd par excellence. By contrast, Franziska valued her clientele of regular, European guests or *Stammgäste*, because with such clients there was scope to communicate more effectively and to forge relationships that lasted over the years. While she struggled to find keen listeners or friends to chat with in the village, some of her guests proved to be great confidants who she was able to share her secrets and pain with, sometimes *thanks* to their 'outsiderness' and the temporary nature of their stays. 'They will be gone by tomorrow, so I can let off steam, it does me good', she once told me after having spent the night discussing her problems in the kitchen with a German guest.

This reveals how, as Valerio Simoni's work has shown, economic and social, intimate interest are far from irreconcilable in tourism encounters, as they can participate in the creation of an exchange based on commonality and reciprocity under certain conditions (Simoni 2014, 2016). To be sure, encounters between native locals and tourists also sometimes resulted in moments of unexpected cross-cultural understanding, as Adrian enjoyed or as another native farmer explained to me. One day as he was working on his farm, he felt he was being observed by Arab tourists who approached him. Thinking they wanted to take a picture of him – and subject him to their 'tourist gaze' – he was surprised when they asked him about how he milked cows and wanted to take pictures of the milking system. One of the men showed him a picture of his own cattle – camels – in the Arabic peninsula. The tourist was genuinely interested in the technicalities of the farmer's job because he was able to relate to them, and vice versa.

This was different for Franziska, who explained that, if none of these outsider, but culturally proximate, guests (Swiss or Westerners) were booking her holiday rental, she would prefer to close her rental business completely, rather than host Asian or Arab guests. 'I do not need them', she said. I could not help to see the contradiction, after she had told me repeatedly about the absolute necessity of the cable car for the future of the valley, and of her children, precisely because it was to secure the future, by attracting these guests. Like all locals, Franziska knew that JR, winter tourism and her family's subsistence were all upheld and cross-financed by the presence of these different, new visitors. What she expressed, however, was that *she* did not want to host them.

TOURISM RESISTANCE BEYOND THE POLITICS OF THE WEAK

These accounts reveal the increasing difficulties of locals, the toured ones, especially of native background, to find meaning and comfort in a business without which they nonetheless claimed they would be 'nothing'. Although tourism holds the promise of securing a future for those who would be deprived of it otherwise, the experiences of locals like Franziska, Thomas, Tobias and Moritz highlight what seem to be some increasingly upsetting consequences of tourism. All spoke of these problems from their native point of view, by describing a business that increasingly did not 'profit' the locals, especially *Einheimische*.

Critical accounts of tourism, casting the negative effects of tourism on native communities, underline this problem, too. Various scholars have described how indigenous populations of the Global South in particular have, until now, experienced dispossession, as they have been subjected to an intrusive 'tourist gaze' and undergone the commodification of their bodies, traditions, history and environment while working in the industry under exploitative working conditions (Nuñez 1963; Ryan and Aicken 2005; Azcárate 2020; Mostafanezhad, Azcárate and Norum 2021). If particularly present in postcolonial contexts, tourism's subordinating power has shaped the experience of locals of Western destinations too: Boissevain describes inhabitants of the latter as the 'weak' because they 'depend on the goodwill of foreign visitors', who are the 'powerful' (Boissevain 1996: 14). The dependency on tourism appears as undoubtedly more problematic for those who live off it than for those who rely on it to find an escape from the everyday for a couple of weeks a year.

In this context, various works recount the strategies of resistance and adaptation mobilized by dependents to 'cope' with tourists, often acknowledging their creativity (e.g. Milano, Novelli and Russo 2024; see Novy and Colomb 2020 for a review). Yet, the reality of tourism dependency they describe remains shaped by the politics of the weak or a 'politicization from below' (Colomb and Novy 2016: 5). Gravari-Barbas and Jacquot, who have worked on movements of resistance against tourism in Paris, speak of 'micro-' – or 'infra-' – politics because of the informal and non-institutional nature of strategies put in place by locals who live with tourism's most nefarious consequences (Gravari-Barbas and Jacquot 2016). These are essentially the 'politics of the dominated', which are often implicit and 'not always intentionally political' (ibid.: 35). Although resistance to urban tourism often tends to encompass and intersect with other struggles,[9] Gravari-Barbas and Jacquot's work speaks of the structural inequality opposing the makers and consumers of tourism to those who live in tourist destinations.

In the Bernese village and many other Swiss destinations, the actions of the more or less explicit contestation of tourism development – like in the play *Alpine Glow* – along with the strategies of division and circumvention, as well as the fencing off of parts of the valley, need to be read against this background.[10] The case of rural tourism points to an even more contested role of this industry, as it tends to develop into monocultures, in contrast with the multisector economies of cities.[11] When, like in my Alpine field site, tourism develops and comes to represent the only alternative to poverty and isolation, it exerts strong pressure on those who depend on it. Under conditions of great competitiveness and of a growing monopoly of the few, tourism (or, rather, a specific form of it) manifests in its most problematic forms. However, I believe that to conclude on domination of the 'toured ones', as I have up until now, fails to grasp the complexity of the story of tourism reliance and its navigation in the Swiss valley, but also in many other European and/or Western contexts and destinations.

Anthropologist Jaro Stacul's findings have illustrated how residents of the Italian Alpine Vanoi Valley expressed strong opposition to outsiders (e.g. urban Italians), whom they perceived as 'intruders' threatening local autonomy and social order (Stacul 2003a). This rejection, Stacul shows, did not relate to the number of tourists, which was very limited when compared to the significantly larger influx seen in neighbouring resorts. Rather, it revealed a localist concern about external encroachment on community life. Such context points to the fact that, in regions politically shaped by localism or nativism like the Vanoi Valley or the Bernese Highlands, critical discourses on tourism cannot *solely* be approached as manifestations of tourism's (undeniable) extractive realities (Azcárate 2020), as the tourism resistance frame of analysis has usually implied.[12] The tale of the dispossessed natives mobilized by people like Moritz or Franziska in my Alpine field site also reflects exclusionary sensibilities that should be acknowledged as such (see Screti 2022 for a similar point). The local reactions and ways of navigating tourism I have recounted here invite a reading beyond tourism resistance as the 'politics of the weak'.

Contrary to locals living off tourism in the Global South, the mostly native and Swiss interlocutors who have featured in the last chapters enjoy greater privilege than many of the guests they host in the Swiss Alps. 'New' guests, in particular, may be travellers (*Reisende*) on whose 'goodwill' the toured ones (*Bereiste*) depend, but they enter a space of discrimination and whiteness when they travel to Europe, and to the Alps in particular (Dickinson 2010). This differs from the positions of bourgeois European travellers, who originally came to the Alps with greater economic, cultural and social capital than the locals. Nowadays, many of the 'historical', returning Western guests of the region stand on a more or less equal footing to those who were born in the valley. But in Switzerland in particular, recent reactions to (over-)tourism and anti-liberal sentiments intersect in important ways that have to be accounted for (see Leitenberg 2025).[13]

At the same time, the valley's political nativism reveals that one cannot completely do away with the category of 'native' either, even if in a European context that significantly differs from that of the indigenous tourism of the Global South. As the jokes about the AI drivers – which play on the absurd juxtaposition of these travellers and the native locals – capture, mass tourists stand in starkest opposition to native locals. In the subsequent pages, I will explore further how attitudes towards so-called mass tourists cannot be divorced from a broader Swiss and European context of political nativism.

ALIENATING GUESTS

Moritz bluntly recounted to me his interaction with one Arab guest who drove up the barred road leading to his house: 'I swore and said in English, "You turn around or I call the police". You know they are very aggressive, especially the Arabs. The Asians apologize but the Arabs say, "What's the problem? What's the problem?" And then the police come and control them, but they don't respect our laws'. On another occasion, he told me: 'The Arabs just do as they please, they drive wherever and park on the sidewalk. If we do that, we have problems. I had my car parked in the wrong spot just now [before the interview] and the police called me and said I had to move the car otherwise I would be fined. If they do that to Arabs, they just don't care. It's a completely different mentality. Women are veiled and men are playboys. They are difficult (*schwierig*) guests'. Moritz's impulsive temperament and attitude were very different from the calm and reasonable character of Franziska, who would never have exposed herself in such a way. Like her, many locals were more cautious in their critiques or gossip about non-Western guests, which they shared with relatives and friends, or, on some occasions, with myself. However, the stereotypes that Moritz used against these tourists – their aggressive or more submissive nature and their unwillingness to adapt – were close to what she, like many locals in the valley and elsewhere in Switzerland (e.g. Eggli 2021), also thought.

This echoes with what others have described of tourist destinations and the common nationality-talk, sometimes accompanied by minor acts of humiliation or racist comments, that shapes the local narratives (Pi-Sunyer 1989; Simoni 2008). In the valley, next to the 'aggressive' Arabs, Indians were said to be difficult because they brought with them their 'caste-thinking system' and mistreated others, Chinese were said to be dirty, and so on. But beyond the differences between nationalities, 'mass tourism' was usually portrayed, as a whole, as a faceless and absurd type of tourism detached from local concerns, in spite of the great disparities among guests that this term seemed to describe. Almost all my interlocutors – native or not – reproached these tourists for their unwillingness to adapt to local rules, local laws, or to 'integrate', as Anna for instance stated. Locals claimed that *they* had to adapt when travelling abroad, while their non-Western guests refused to do so. One local hotelier, for instance, close to the SVP party, told me that he refused to host guests from the Arabic peninsula if the women wore hijabs because this did not 'fit' aesthetically with the Alpine style of a hotel in the Swiss Alps; it was not what others tourists 'expected' and sent a discriminatory message to women. Such portrayals of particular guests' reluctance to adapt to local rules, customs or even aesthetics, because of their 'incompatible' racial, cultural or religious difference, cannot be seen as simple reactions to tourism's growth in the valley.

Markus, one of Beat's friends and fellow native opponent, revealed this when he described non-Western mass tourists and refugees as 'the same': according to him, both simply came and went without any genuine interest in the place, and they even contributed to its degradation. He referred to the youngsters from a refugee centre for unaccompanied minors, mostly Eritrean and Afghan, which had been set up in the valley but had closed in early 2018. Seeing the refugee kids 'hang out' in groups close to the information centre where they had Wi-Fi connection or at the train station, Markus had heard that they were only interested in leaving the valley to go to Interlaken – exactly like the mass tourists that JR brought in, from which 'we *Einheimische* no longer profit'. This confusion between non-white guests and migrants also reflected what I witnessed when spending time with two young boys from Afghanistan, housed in the valley's refugee centre. On multiple occasions, I saw and heard Swiss locals mistake them for Indian tourists, talking about them as such and addressing them in English. As one of the boys told me, he was particularly fond of the valley because he too came from a mountain village in Afghanistan. He was irritated that locals were still treating him like a guest, in spite of the two years he had spent living and working in the valley after he had turned eighteen. This, and the difficulties in finding an accommodation in the village, had led him, like the other refugees of the centre, to leave.

Switzerland's historical fears of cultural alienation (*Überfremdung*) provide an explanation for the correlation made by many of my interlocutors between 'mass tourists' and refugees, and their portrayal as invasive 'masses' of radical others coming into the small village. From the end of the nineteenth century, the threat of being overwhelmed and alienated by foreigners has shaped Swiss everyday narratives but also the country's migration policies, which are aimed at preserving the cultural integrity of a small, wealthy nation in the middle of Europe (Niederberger 2004). I will describe the formation of this idea in more detail in the next chapter in relation to a migrant workforce, but for now it suffices to say that those who fear *Überfremdung* view foreigner presence generally as estranging for those who 'naturally' inhabit Switzerland. In this line, Markus for instance believed that 'Switzerland was the smallest corner of the world' threatened by overpopulation, notably because of migration, which aligned with many criticisms of the valley's saturation with tourists. Hans, another native local, told me he would be 'green' because he was attached to the landscape and the integrity of small Switzerland, but the leftist positions of the latter prevented him from voting for them. 'If they really were green, if they wanted the landscape, they would have to be against immigration, against the fact that more and more people migrate to the country!'

Such discourses of infiltration have primarily targeted migrant presence, and have recently revolved around the figure of the non-white refugee and the Muslim migrant (Michel 2015; dos Santos Pinto et al. 2022). Yet, the figure of the tourist also surprisingly appears in historical documents that framed national concerns about overforeignization. In a document entitled 'The problem of the foreign workforce' issued by the BIGA (Federal Office of Trade and Labour) in 1964, the large number of tourists at the time was described as impacting on the 'ways of speaking, sometimes also the customs, on public transport, in hotels, restaurants, shops', which had the consequence of eroding the 'informal-democratic (*familiär-demokratische*) character [of places], which is a basis of the Swiss characteristic (*Eigenart*)' (BIGA 1964: 135, cited in Inderbitzin 2002: 218).

This quote eerily echoes the statement of the hotelier I cited in the previous chapter, who explicitly pointed out the new guests' disruption of the 'familiar atmosphere' of historical hotels. Franziska's irritation towards Asian tourists who asked for water for free and unexpectedly showed up in an area usually reserved for Swiss locals and Western tourists follows a similar logic. This was an elevated space of greater equality, where hosts and (European) guests addressed each other with informal pronouns (*Du*), which the Asians disrupted. Instead of seeing tourist presence as erosive as such, it is the 'mass' or non-Western tourists – be they travelling in groups or individually – that are perceived as alienating. Importantly, some Western tourists themselves were engaged in racializing, exclusionary talk on other guests, like one German woman commenting on a Facebook post that 'gondolas with announcements in Arabic are simply unacceptable'. The newness and difference of Arab guests, this woman implied, clashed with what one could aesthetically expect from an Alpine valley, and therefore was bothersome when travelling through it.

OUT OF PLACE

One extremely common complaint that accompanied critiques of the phenomenon of 'mass tourism' in local narratives was the association between non-Western guests and the damage, littering and waste they left behind, to be cleaned by the hosts. Countless stories of overflowing bathrooms, smell and chaos were retold by locals, often, but not always, indirectly. Pictograms had been installed in the valley's public toilets too, indicating the appropriate way to use the bathroom for non-European guests (by having one's feet on the ground and not *on* the toilet) because of previous cases of damage and degradation of public washrooms. Various locals said they had experienced cleaning for hours after hosting Arab or Asian guests because of their unfamiliarity with the European infrastructure. One hotelier also told me that hotel rooms were often left extremely untidy and disrupted when rented out to non-Western guests, and reported that his team had found traces of meals cooked in hotel rooms and of overflowing bathtubs and toilets. He showed me information sheets hanging in his hotel, explaining the correct use of bathrooms and informing guests in Arabic that they would be charged additional fees in cases of water damage. Other signs were displayed in the region to remind tourists in English and Arabic to 'keep Switzerland clean and take your rubbish with you' and help 'preserve our beautiful nature'.

In some cases, references to dirt were not material or visible, as I witnessed on one summer's day in 2018. One particular group of young men from Saudi Arabia, who had travelled through Switzerland with their rental car, an Audi with a VD number plate, parked next to the village's train station and left their engine running. A local employee of the railway company, who had refused to help them when they searched for help moments earlier, yelled at them, in English: 'You have to turn off the car. I know in your country you don't, but here we have nice fresh air!' Pollution here also comprised another type of waste, left behind by these tourists, who were presented as environmentally unaware, contrary to the Swiss and their great respect for nature – a respect that was not the result of a political commitment or 'trend' from the greens, but rather a form of Alpine lifestyle inherited from the past.[14] The fact that Switzerland and tourism importantly contributed to the production of material and environmental waste was disregarded.[15]

Anthropologist of tourism Valerio Simoni speaks of hospitality as a 'moral order' and 'a measure of control', which is aimed at 'socializing, containing and controlling strangers' (Simoni 2018: 301). Living off tourism and the variety of guests one needs to host, he shows, is in itself a contentious and contested matter. I do not wish to minimize the challenges of having to cope with tourists who have other customs, which may easily transform into misunderstandings and confrontations. However, many locals' insistence on the untidiness or dirtiness of many guests often served to represent a counterpoint to Switzerland's cleanliness, beauty and purity. This has deeper, social implications that require an attentive examination of what being accused of littering, overflowing, damaging – of spreading dirt, in other words – means.

Mary Douglas famously described how dirt does not stand for a universal, objective reality but instead points to a form of disorder, whose perception is 'in the eye of the beholder' (Douglas 2003: 2). Douglas's work showed how 'ideas about separating, demarcating and punishing transgressions have as their main function to impose system on an inherently untidy experience' (ibid.: 3). Dirt is what transgresses boundaries and rules to become what Douglas coined as 'matter out of place' and, in this context, 'some pollutions are used as analogies for expressing a general view of social order' (ibid.).

Examples of people who have been associated with dirt abound, in Switzerland and beyond: the poor, migrants, slaves, racialized people (Zimring 2015; Stallone 2022). In the Alpine valley, the Afghan and Eritrean refugees and the personnel who worked with them in the valley's accommodation centre told me about the often baseless and relentless accusations of degradation of the valley's infrastructure they received. For Laura, a former centre worker, these complaints highlighted the racism of some of the refugee centre's neighbours towards migrants, who were simply trying to live their lives. Like for the 'mass tourists', the dirtiness they represented was not just a matter of behaviour but also often entailed implications of cultural, visual and racial incompatibility with the Swiss values of order and purity (Mottier 2000).[16]

One important difference between refugees and the new guests coming to the valley was nonetheless the latter's undeniable economic capital, which was sometimes greater than that of the Swiss locals. While Franziska could avoid hosting 'mass tourists', other locals I met explained they could not afford to give up on accommodating even the most complicated guests, who made for an increasingly large proportion of guests in the valley (see Chapter 3). This was the case for Nina and Paul, two retired locals who owned an old house on one of the valley's sides. Next to his impressive collection of local objects and pictures, Paul had printed a sheet of paper with pictograms on how to use the toilets. When I expressed my surprise at seeing this, he explained with a smile that this was unfortunately the way it was now with the new tourists: one had to explain even the most basic rules of decency. The pictograms were accompanied by very precise descriptions of what and what not to do when on the toilet, which seemed to reflect Paul's humour. Paul and Nina exchanged stories about guests, especially the Arab ones and the trails of bad memories they left, which Nina admitted as being more memorable than when things went smoothly.

They recalled one guest in particular. A couple of weeks before my visit, an Arab couple did not want to pay for damages after they had caused the bathroom of Nina and Paul's rental flat to overflow. Paul explained that all this was related to some religious customs according to which Arab guests were forbidden to use the same shower to wash and that, for this reason, they probably used the sink. After the guests stubbornly refused to pay, Nina and Paul called the police, who informed the guests they were obliged to pay. Faced with their reluctance, the policeman called the ambassador of these guests' country, who then told Paul and Nina to 'please have compassion for these poor people'. Nina and Paul were outraged. 'The Indians and the Arabs who come here are very wealthy, they go to shop in expensive, luxurious shops in the valley', Nina said to me. The policeman then took the guests' passports and asked them to pay.

I believe that Nina and Paul told me this story to highlight the pressure exerted by this type of tourism on locals in the valley, as a form of pressure on dependents who had no other choice but to follow the rules of the game. But the way they presented it, interspersed with remarks on their guest's odd, reprehensible customs or intolerance, made the Arab tourists into 'others' they thought were given the unfair, monetary power to dominate them. This domination was limited: confronted with the policeman, who Paul said even 'illegally' took the passports to encourage these guests to pay, the tourists were made aware of the rules. Paul was visibly glad to count on the policeman's complicity. This example further reveals that if the non-Western tourists, especially Arabs, had financial resources, it did not make them more welcome; for locals, their money was a subjectifying and alienating power – an idea that was implicit in many of my talks with locals on the subject.[17]

CONCLUSION

If puzzling at first, Franziska's accounts of the need to go forward, to invest in and secure the future for her children, and her personal yearnings for the re-emergence of values of the (romanticized) past, point to the intrinsic instability of tourism as both a forward- *and* a backward-looking enterprise. In the Alpine valley, like elsewhere, this tension has been addressed and navigated by dependents who have regularly complained of tourism's excesses while trying to carve out the most meaningful existences in an imperfect situation. Yet, the sense of increasing fragility of tourism dependency that emerges from this chapter reveals the need to adopt more subtle analysis.

The exponential rate of tourism development since the 1990s has had material consequences for many locals. The navigation strategies of someone like

Beat, whose story I have retold in the second chapter, is tragically emblematic of this reality. During the day and in the high season, he explained working as being confronted by the disturbing 'masses' as a JR employee, but at least he could return to his home on the valley's sides, his *Heimat*, to find quietness and beauty. Since the renovation programme and expansion of the tourist infrastructure, with the cable car project, Beat explained his worries of being subjected to the penetrating gaze of guests travelling on the cable cars, who would be able to look in on his house, his intimacy, his privacy. The passing of the cabins in front of his house in summer signified it would be traversed by a shadow, every thirty seconds, he told me. At the same time, he was already suffering on account of his complete exclusion and humiliation in the village, as the 'enemy of tourism'. The strategies of divide and retreat that Beat put in place had slowly collapsed, as his *Heimat* shrunk in the face of infrastructural sprawl. Although in more minor ways, people like Franziska, who lived in a green belt, were also impacted by the dimensions and sacrifices imposed by the new project. The ways she, as a tourism dependent, navigated tourism's paradoxes by spatially defining places of preservation and innovation had become increasingly complicated, with new investments the valley had to make to secure a competitive future – a global reality of tourism as an exponentially growing, environmentally and socially extractive, 'predatory' industry (Azcárate 2020; Mostafanezhad, Azcárate and Norum 2021).

At the same time, the perception of tourism's disruption and dispossession, its increasing lack of 'profitability' or value for locals, particularly prominent among native villagers, cannot be solely attributed to the capitalist power of tourism acting on helpless dependents. Claims on the decreasing profit derived from tourism often emerged from a nativist point of view, where ideals of social order regularly took on racializing and culturalizing undertones. If, as Peter Geschiere has put it, nativism denotes an incapacity to deal with history and the transformations it implies (Geschiere 2009), it is easy to see how new guests always appear to be the threatening 'other' to those emplaced naturally. Some formerly despised guests had however become relatively accepted in the valley, such as the Japanese, because they had been largely replaced by even more alienating others. At the same time, the figure of the mass tourist seemed to have changed, at least for some, when I came to the valley shortly after the 2020 pandemic. One of my local interlocutors stated that 'mass tourism' during the pandemic in fact consisted of Swiss lowlanders on mountain bikes, who crossed through the fields without any effort to adapt to local rules. This picture of mass tourism was soon replaced by a re-emerging concern over the non-Western tourists, but this reveals the intrinsically problematic relationship with a changing 'other' in nativism.

Between nativism and capitalism, this chapter has accounted for the increasingly difficult navigation of tourism ambiguities for locals via strategies of divide, retreat, circumvention, avoidance and the hierarchization of places, processes and guests. The realm of the future – pressuring and disorientating – increasingly invaded the domain of the past – comforting, authentic and pure – when local strategies previously allowed to maintain such realms separately. While 'living where others spent their holidays' was a privilege, as Franziska had acknowledged, it was also a deeply and increasingly difficult, dissatisfying and slippery reality.

NOTES

1. As I have shown in the last chapter, even supporters of the cable car were critical of the latest tourism developments of the valley (see Chapter 3).
2. The concerns on the *Verbetonierung* of Switzerland gather environmentalists and conservative politicians; see the example of Franz Weber, the initiator of the Second Home regulation (*Zweitwohnunggesetz*, ZWG) (see SaW Redaktion 2012). See Chapter 1, footnote 7.
3. In German, the terms *heimelig* (homely) and *unheimlich* (uncanny) both derive from the idea of *Heimat*.
4. This is, however, an extremely gendered activity: the *Feierabend* was mostly a male activity.
5. In local talk, however, 'tourism' and 'capitalism' remained overwhelmingly positive categories in a region (and a country) that, in many ways, epitomized success and financial stability. To distance oneself critically from capitalism, or even to name it as such, was highly improbable because it was the prerogative of the left, the socialists or the communists – all insults in this part of Switzerland (see Chapter 3).
6. A newspaper article included the following: 'When Arab tourists drive, they usually drive quite badly. They also seem to "overlook" the new driving bans' (Karlen 2014a, b; Lehmann 2018).
7. See the article entitled 'The marvellous world of AI-Drivers' (Abegglen 2023).
8. The new awareness campaign, as well as media reporting and a Facebook thread on the topic, all dating to summer 2024, also point to this increasing problematization of non-European tourists as invaders in the whole region (e.g. Aschwanden 2024).
9. Associated with struggles for the 'right to the city', for instance (Novy and Colomb 2020; Milano, Novelli and Russo 2024; Santos, Malet Calvo and Nofre 2025).
10. Colomb and Novy's exhaustive list of points of contention related to urban tourism are summarized under three themes, which apply to the case of my field site too: 'The negative effects and externalities caused by tourism on people and places . . .; equity impacts, i.e. the distribution of the costs and benefits of urban tourism among various groups and spaces – often very unevenly spread; the politics of urban tourism, including especially what is perceived as a skewed political agenda in favour of the 'visitor class', and the prioritization of development for exchange value over development for use value in the definition and conduct of public policies' (Colomb and Novy 2016: 5).

11. On the presence of companies who have an economic monopoly in the Alps, see Bätzing 2015: 197–98.
12. The complicated levels of critique of tourism and their illiberal tendencies have been discussed in the case of Goa (Routledge 2001; Saldanha 2002), as well as the Taiwanese context (Garrett 2016), but remain underexplored ethnographically in a European, Western context. In Europe, some have pointed to the individual, self-serving claims that stand behind movements opposing (over-)tourism with cases of Nimbyism-reported tourism resistance literature (Colomb and Novy 2016: 16; Litvin, Smith and McEwen 2020). My argument here, however, goes in a different direction.
13. For how tourism and right-wing populism or racism shapes images of Swiss destinations, see Feighery 2015.
14. This problem also recently featured in the new local tourism awareness campaign, instructing tourists to switch off the engines of their cars when stopping. Not doing it could result in fining, it specified.
15. https://www.swissinfo.ch/eng/climate-change/swiss-co2-emissions-small-country-big-footprint/45810036#:~:text=This%20content%20was%20published%20on%20Jun%209%2C%202020%20Each%20person,recycles%20about%20half%20of%20it. (last accessed 17 April 2025).
16. Similar criticisms were directed at Ukrainian refugees hosted in the valley, with some locals reproaching them for their alcohol consumption and the negative image this allegedly cast on the village. Despite these criticisms, support for Ukrainian refugees was notably stronger than for those from the centre closed in 2018, with the local population organizing initiatives such as German language courses to aid their integration. This support was driven in part by the efforts of newer residents, and underpinned by a sense of humanitarianism deeply rooted in shared Christian values between some locals and the Ukrainian refugees.
17. This was reinforced during the FIFA World Cup in Qatar in 2022, when a global context of contestation and criticism of the country for its lack of respect for human rights was also interspersed with racist, Islamophobic arguments (Fernández 2022).

REFERENCES

Abegglen, Ben. 2023. 'Die Wundersame Welt Der AI-Fahrer'. Retrieved 15 April 2025 from https://www.plattformj.ch/Artikel/209901/.

Aschwanden, Erich. 2024. 'Swiss Locals Are Increasingly Irritated by Throngs of Tourists', *Neue Zürcher Zeitung*, 17 July 2024. Retrieved 15 April 2025 from https://www.nzz.ch/english/swiss-locals-are-increasingly-irritated-by-throngs-of-tourists-ld.1838320.

Azcárate, Matilde Córdoba. 2020. *Stuck with Tourism: Space, Power and Labor in Contemporary Yucatan*. Oakland: University of California Press.

Bätzing, Werner. 2015. *Die Alpen: Geschichte und Zukunft einer Europäischen Kulturlandschaft*. Munich: C.H. Beck.

Boissevain, Jeremy. 1996. *Coping with Tourists: European Reactions to Mass Tourism*, Vol. 1. New York: Berghahn Books.

Boissevain, J., and Tom Selwyn. 2004. *Contesting the Foreshore*. Amsterdam: Amsterdam University Press.

Cole, John W., and Eric R. Wolf. 1999. *The Hidden Frontier: Ecology and Ethnicity in an Alpine Valley*. Los Angeles: University of California Press.

Colomb, Claire, and Johannes Novy. 2016. *Protest and Resistance in the Tourist City*. Abingdon: Routledge.

Dickinson, Edward. 2010. 'Altitude and Whiteness: Germanizing the Alps and Alpinizing the Germans, 1875–1935', *German Studies Review* 33(3): 579–602.

Dos Santos Pinto, Jovita, et al. 2022. *Un/Doing Race: Racialisation en Suisse*. Zurich: Seismo.

Douglas, Mary. 2003. *Purity and Danger: An Analysis of Concepts of Pollution and Taboo*. Abingdon: Routledge.

Eggli, Florian. 2021. 'Living with Tourism in Lucerne: How People Inhabit a Tourist Place'. PhD dissertation, Université de Lausanne.

Feighery, William. 2015. 'Right-Wing Visual Rhetoric and Switzerland's Tourism Image', in *Tourism Research Frontiers: Beyond the Boundaries of Knowledge*. Bingley, UK: Emerald Group Publishing, pp. 105–22.

Fernández, Belén. 2022. 'The Massive Hypocrisy of the West's World Cup "Concerns"', *Al Jazeera*, 28 November 2022. Retrieved 17 April 2025 from https://www.aljazeera.com/Opinions/2022/11/28/the-Massive-Hypocrisy-of-the-Wests-World-Cup-Concerns.

Garrett, Daniel. 2016. 'Contesting China's Tourism Wave: Identity Politics, Protest and the Rise of the Hongkonger City State Movement', in Claire Colomb and Johannes Novy (eds), *Protest and Resistance in the Tourist City*. Abingdon: Routledge, pp. 121–42.

Geschiere, Peter. 2009. *The Perils of Belonging: Autochthony, Citizenship and Exclusion in Africa and Europe*. Chicago: University of Chicago Press.

Gössling, Stefan, Scott McCabe and Ning Chen. 2020. 'A Socio-Psychological Conceptualisation of Overtourism', *Annals of Tourism Research* 84: 102976. doi: https://doi.org/10.1016/J.Annals.2020.102976.

Gravari-Barbas, Maria, and Sébastien Jacquot. 2016. 'No Conflict? Discourses and Management of Tourism-Related Tensions in Paris', in Claire Colomb and Johannes Novy (eds), *Protest and Resistance in the Tourist City*. Abingdon: Routledge, pp. 45–65.

Inderbitzin, Ivan. 2002. 'The Foreignisation Process in Switzerland: The Swiss and Their Ausländer'. Phd doctoral thesis, Anthropology Department, Monash University, Melbourne, VIC.

Jaeggi, Rahel. 2014. *Alienation*. New York: Columbia University Press.

Karlen, Alex. 2014a. 'Araber in Interlaken – Bussen, Bussen, Bussen', *Tages Anzeiger*, 2 September. Retrieved 15 April 2025 from https://www.tagesanzeiger.ch/araber-in-interlaken-bussen-bussen-bussen-890830873811.

———. 2014b. 'Schlechte Autofahrer Auf Abwegen.' *Berner Zeitung*, 1 September. Retrieved 15 April 2025 from https://www.bernerzeitung.ch/Schlechte-Autofahrer-Auf-Abwegen-505996928803.

Lehmann, Fritz. 2018. 'Wie Behörden Gäste und Einheimische erziehen wollen', *Berner Zeitung*, 13 September. Retrieved 15 April 2025 from https://www.bernerzeitung.ch/Wie-Behoerden-Gaeste-Und-Einheimische-Erziehen-Wollen-430078933031.

Leimgruber, Walter. 2019. 'Constructing a Home: Heimat as Expression of Privilege, Belonging, Exclusion and Identity', *Journal for European Ethnology and Cultural Analysis* 1: 142–67.

Leitenberg, Danaé. 2018. 'Stranger in Paradise', *Allegra Laboratory*. Retrieved 15 April 2025 from https://allegralaboratory.net/stranger-in-paradise-displacement/.

Leitenberg, Danaé. 2025. 'The Business of Foreigners: Ordering Mobilities in the Touristic Swiss Alps' in Viviane Cretton and Andrea Boscoboinik (eds), *Mountainscapes: Pathways and Mobilities in the Alps and Beyond*. New York: Berghahn, pp. 38–55.

Litvin, Stephen W., Wayne W. Smith and William R. McEwen. 2020. 'Not in My Backyard: Personal Politics and Resident Attitudes toward Tourism', *Journal of Travel Research* 59(4): 674–85. Doi: 10.1177/0047287519853039.

Michel, Noémi. 2015. 'Sheepology: The Postcolonial Politics of Raceless Racism in Switzerland', *Postcolonial Studies* 18(4): 410–26.

Milano, Claudio, Marina Novelli and Antonio Paolo Russo. 2024. 'Anti-Tourism Activism and the Inconvenient Truths about Mass Tourism, Touristification and Overtourism', *Tourism Geographies* 26(8): 1313–37. Doi: 10.1080/14616688.2024.2391388.

Mostafanezhad, Mary, Matilde Córdoba Azcárate and Roger Norum. 2021. *Tourism Geopolitics: Assemblages of Infrastructure, Affect and Imagination*. Tucson: University of Arizona Press.

Mottier, Véronique. 2000. 'Narratives of National Identity: Sexuality, Race and the Swiss Dream of Order', *Revue Suisse de Sociologie* 26(3): 533–58.

Niederberger, Josef M. 2004. *Ausgrenzen, Assimilieren, Integrieren: Die Entwicklung einer schweizerischen Integrationspolitik*. Zurich: Seismo.

Norum, Roger, and Mary Mostafanezhad. 2016. 'A Chronopolitics of Tourism', *Geoforum* 77: 157–60.

Novy, Johannes and Colomb, Claire. 2020. 'Overdosed, Underplanned or What? Making Sense of Urban Tourism's "Politicisation from Below"', in Oskam, Jeoren A. (ed.), *The Overtourism Debate*. Leeds: Emerald Publishing Limited, pp. 75–94.

Nuñez, Theron A. 1963. 'Tourism, Tradition and Acculturation: Weekendismo in a Mexican Village', *Ethnology* 2(3): 347–52.

Petite, Mathieu. 2013. 'Les montagnards face aux écolos. Une énième déclinaison au travers de l'initiative Weber', *Journal of Alpine Research* (Hors-Série).

Pi-Sunyer, Oriol. 1989. 'Changing Perceptions of Tourism and Tourists in a Catalan Resort Town', in Valene L. Smith (ed.), *Hosts and Guests: The Anthropology of Tourism*. Philadelphia: University of Pennsylvania Press, pp. 187–99.

Routledge, Paul. 2001. '"Selling the Rain", Resisting the Sale: Resistant Identities and the Conflict over Tourism in Goa', *Social & Cultural Geography* 2(2): 221–40. Doi: 10.1080/14649360120047823.

Ryan, Chris, and Michelle Aicken. 2005. *Indigenous Tourism: The Commodification and Management of Culture*. London: Taylor & Francis.

Saldanha, Arun. 2002. 'Identity, Spatiality and Post-Colonial Resistance: Geographies of the Tourism Critique in Goa', *Current Issues in Tourism* 5(2): 94–111. Doi: 10.1080/13683500208667910.

SaW Redaktion. 2012. 'Umweltschützer Franz Weber: "Stoppt die Einwanderung!"', *Thurgauer Zeitung*, 21 July. Retrieved 15 April 2025 from https://www.thurgauerzeitung.ch/verschiedenes/umweltschutzer-franz-weber-stoppt-die-einwanderung-ld.1618622.

Santos, Priscilla, Malet Calvo, Daniel, and Nofre, Jordi (2025). 'Navigating between Resistance and Unintentional Collaboration: The Role of Left-wing Grassroots Associations in the Tourist City'. *Urban Studies* 62(4): 737–53.

Screti, Francesco. 2022. 'Populism in Mediated Anti-Tourism Discourse: A Critical Analysis of the Documentary Tourist Go Home!' *Journal of Tourism and Cultural Change* 20(5): 617–32. Doi: 10.1080/14766825.2021.1966024.

Simoni, Valerio. 2008. 'Riding Diversity: Cubans'/Jineteros' Uses of 'Nationality-Talks' in the Realm of Their Informal Encounters with Tourists', in Peter M. Burns and Marina Novelli (eds), *Tourism Development: Growth, Myths and Inequalities*. Wallingford, UK: CABI, pp. 68–84.

———. 2014. 'From Tourist to Person: The Value of Intimacy in Touristic Cuba', *Journal of Tourism and Cultural Change* 12(3): 280–92.

———. 2016. *Tourism and Informal Encounters in Cuba*, Vol. 38. New York: Berghahn Books.

———. 2018. 'Business, Hospitality and Change in Cuba's Private Tourism Sector: A View from Casas Particulares in Viñales', *Tourism Planning & Development* 15(3): 293–312. Doi: 10.1080/21568316.2017.1399436.

Stacul, Jaro. 2003a. *The Bounded Field: Localism and Local Identity in an Italian Alpine Valley*, Vol. 18. New York: Berghahn Books.

———. 2003b. 'The Privatisation of Locality: Integralist Political Engagements in Northern Italy at the Turn of the Millennium', *Cambridge Journal of Anthropology* 23(2): 36–53.

———. 2006. 'Neo-Nationalism or Neo-Localism? Integralist Political Engagements in Italy at the Turn of the Millennium', in Marcus Banks and Andre Gingrich (eds), *Neo-Nationalism in Europe and Beyond: Perspectives from Social Anthropology*. New York: Berghahn Books, pp. 162–76.

Stallone, Sabrina. 2022. 'Smart Complaints? 'Wild' Things and the Promise of (Un)Happiness in Urban Crowdsourcing Apps', *Urban Transcripts Journal* 5(2).

Vigh, Henrik. 2009. 'Motion Squared: A Second Look at the Concept of Social Navigation', *Anthropological Theory* 9(4): 419–38. Doi: 10.1177/1463499609356044.

Zimring, Carl A. 2015. *Clean and White: A History of Environmental Racism in the United States*. New York: NYU Press.

CHAPTER

5

NEVER LIKE THEM

'We are invisible', said Ines, a former chambermaid hailing from Portugal, in her fifties, and now working in a sports shop, as we were walking along the busy main street of the village. At other times, she would state, 'We are just like slaves' or 'We did not have any other choice'. Speaking in the name of her fellow hospitality workers – all migrants – in hotels and restaurants, Ines uttered these sentences calmly, with a hint of anger but also a lot of resignation; they were the result of reflections she had been ruminating over the years.

Ines and I met by chance in the shop, when I heard her speaking French to a customer. For weeks, I came to her workplace, where she and I would chat for a minute or two. She enjoyed speaking the French that she had learnt at school, and insisted on finding the right word instead of switching to a Swiss-German language that she now spoke fluently. Ines did not behave with the initial suspicion I encountered when talking to many (migrant) workers – on the contrary. When I finally asked her whether she would be willing to participate in a recorded interview, she responded: 'I have been waiting for you to ask that for weeks. I have a lot to tell'.

Although Ines had managed to leave the job in favour of what many saw as more 'normal' positions – with secure working conditions, regular hours and Swiss co-workers – the reality of what her life as a hotel worker had entailed until a couple of years ago still shaped much of how she thought and who she was. She had now spent most of her existence in the valley, having arrived from a rural village of northern Portugal more than thirty years ago.

Leaving poverty and her parents behind to go to Switzerland, Ines joined her brother, who was working in a hotel in central Switzerland and he found her a job as an 'office' worker – 'Nothing to do with an office, just working in the kitchen', she said. At just eighteen years old, Ines worked in 'very badly paid [conditions] but still better than in my country', and she learnt Italian quickly so that she could communicate with colleagues, who were all Italian. She then moved to the Alpine village upon her brother's advice, without knowing a word of German.

Most of Ines's relatives, along with her neighbours and compatriots, took the same path before and after her, escaping poverty and the lack of prospects after decades of the Salazar dictatorship (Fibbi et al. 2010). Her father refused to see her leave for the next city in Portugal, even for weekends, but working in Switzerland was an acceptable path. 'I never understood that', said Ines. Only one of her seven siblings had never worked in Switzerland, and sometimes she thought that he was the luckiest. He had it hard in some sense but 'at least he is at home', she said. Unlike her, he did not have to face the harsh reality of migration, of having to learn a new language, of having to work hard for little money and of being treated 'like a slave' in Switzerland.

Thinking of her young self arriving in the valley, Ines was now struck by her own naivety. Her then dreams and aspirations of independence contrasted with what was about to happen with her life: some moments of happiness but also a lot of struggles, hardship and sadness. Ines explained that for nearly thirty years she had led a life where the problems of family, relationship and work were completely interwoven: young children, exploitative work conditions, and a controlling husband (who was also a co-worker) made for an explosive cocktail. As much as she tried to maintain a healthy family life and to distance it from the work routine, her husband brought work problems into their home, and the direct proximity with the hotel weighed on her. 'I did not have a life. Work and family were my life.' After a bout of depression, health issues started to manifest, all stress related. The pain triggered by these events was still everywhere to be felt, although Ines had now reached some precarious peace. Alone in her house, she was finally able to do things for herself, like baking, knitting and crafting, surrounded by the pictures of Cristiano Ronaldo and Portuguese flags that her now adult sons had pinned on the walls when they were kids.

Aware of the many injustices she had experienced, Ines thought about writing a book. Her life was made up of the stuff that those she called 'they' – the Swiss and the native villagers – avoided: her house was located in the overcrowded centre, and her job was carried out among tourists 'they' were reluctant to serve. Ines explained that she felt different to the rest of the 'genuine' villagers, who had been born and raised in the valley. Despite spending more than three decades there, mastering the local dialect and then moving out of the backstage of the hotel industry to work on the front lines as a salesperson, she knew she still did not belong to the village. She was not like its native inhabitants. This reality made her doubt her legitimacy to tell her story. 'What is so special and interesting about the story of a foreigner in this place?', she wondered out loud when I told her I was interested in hearing what she had to say about the village. Migrants like her had been given the denigrating, arduous jobs that 'nobody else wanted to do', but they came to Switzerland to work and so should not have expected anything else. Ines often implied that feeling invisible and exploited only revealed unrealistic expectations in a context of repeated forms of injustice against the poor.

The gap between Ines's world and that of the Swiss or native locals that I have described previously in this book became particularly obvious to me when I questioned her on issues that concerned the valley dwellers' common feature, namely their reliance on tourism and the debates around its development. When the village was shaken to its core by the conflicts around the new cable car, Ines did not go to vote. Even if she was legally allowed to do so by then as a 'naturalized' Swiss person who had recently acquired citizenship, she did not position herself in favour or against. 'They do what they want anyway', she said distantly when we talked in 2017 about what seemed to upset a multitude of vocal locals in the media, in assemblies, in the street.

And the play *Alpine Glow* aroused no more interest from her. Like most of my migrant interlocutors, Ines had somehow heard about the play that told the history of the touristification of the village, but did not attend it. Tickets were too expensive and she was busy working long hours as it was summertime, in the middle of high season. Knowing that one of her former employers, the owner of a historical hotel, was deeply involved in the process of creating and financing the play, I could not help but feel that she had contributed to the making of the play thanks to her 'underpaid' job, as she described it. But between the local, native hosts and the European tourists, there was no mention of her kind of position as an outsider *and* a host in the play. When I commented on this, Ines responded, 'We are not part of history, we are in the shadow'. Then she continued, 'I know that when I leave [for Portugal for retirement], I will not be part of history'. The story retold in the numerous history books on the valley seemed to confirm her prevision. Although a member of the biggest migrant community[1] in the village (more than 370 Portuguese out of the 1,129 foreigners living in the village as of 2020), Ines did not appear to be concerned by the history of the place as it was presented, nor was she interested in being involved in the debates around its future.

Bringing the experience of hospitality workers like Ines to the fore, this chapter and the next one stand in stark contrast to those that precede them. I have, until here, mainly focused on Swiss and/or native locals and their experience of tourism dependency, made up of a slippery and increasingly dissatisfying state between the pressures imposed by the making of the future and the preservation of the past. But Ines's apparent disinterest in the temporal narratives and experiences that shaped local understandings of tourism reliance, although herself particularly affected by her activities as a migrant hospitality worker, pointed to another realm of dependence. Following her perspective and that of other migrant workers, I propose to take a step back from the normative understandings of tourism dependency – i.e. what forms of it are deemed valuable or undesirable – to examine the exclusionary nature of the discourses and experiences that sustain them. Here, I seek to interrogate the emergence of such feelings of indifference that Ines, like other migrants, experienced in the valley, in spite of standing in the centre of many touristic transactions. To do so, I trace the historical inclusion of migrant workers in the Alps and in Switzerland under particularly exploitative conditions (see Leitenberg, forthcoming). The subsequent pages explore how nativism shapes another facet of tourism besides the hierarchization of guests according to their background and hospitability: they reveal how it also importantly shapes how labour is distributed among host populations.

A WIN-WIN SITUATION

In the Alpine valley, as in the rest of Switzerland, everyone was aware of the strong reliance on migrant labour and the presence of people like Ines. Some problematized this presence but, for most, these workers did not evoke much more than a distant, global reality of interconnection. As native local Franziska noted, even though she perceived the changes in population in the centre of the village as alienating, the multiculturalism she witnessed in the village came 'from the tourists but not from the workers', unlike in the lowlands. The valley had a proportion of foreigners similar to many urban regions, but the reality of migration was not one that fundamentally shaped the culture of the valley, she stated. My discussions with Swiss villagers seemed to confirm this view: the latter would regularly speak of the 'hosts' or the 'toured ones', as opposed to the tourists or the 'travellers', a binary from which migrants, like other 'intermediaries' that are 'neither hosts nor guests in any conventional manner' (Chambers 1997: 6), were easily erased.

When I brought up the topic of migrant hospitality work, my interlocutors often explained that their presence was the result of the simple, universal rules of the market, made up of exchanges. This was what one of the village's tourism officials explained when I questioned him about the importance of migrant labour in the valley: 'I think that we Swiss are served well (*wir Schweizer sind gut bedient*) because we can count on this [foreign] workforce. On the other side, it [work in Switzerland] helps these people to go back to their countries with good wages and to make a living on their own there (*sich selber eine Existenz dort aufbauen*). For me, this is a win–win situation'.

Following this liberal rationale, common to many Western countries and shared by many Swiss, the simple arrival of migrants on the country's soil marks an improvement in their personal quality of life relative to their lives in their home countries. Media, political and everyday discourses have regularly described Switzerland as a migrant *Eldorado*, a place of security, wealth, democracy and purity.[2] In this view – to which the tourism official seemed to subscribe – outsiders automatically 'win' by entering the Swiss national space. At the same time, migrant presence and especially labour appears as beneficial for the host country. As Dauvergne and Marsden note, such views can even be described as those of a 'win–win–win situation', a virtuous circle equally benefiting destination states, labourers and incoming states (Dauvergne and Marsden 2014).

Ines, however, pointed to a clearly winning party: her employers and the Swiss, broadly speaking, had driven profits up over decades, while workers like her moved in a field over which they had little power or influence. The Swiss were indeed well *served*, not only because the majority[3] of service work was undertaken by migrants but also because they gained important benefits from this exploitative situation. Ines did not envy native locals, who had their own difficulties, but they were *at home*, not only in their own culture and networks but also in their properties.[4] Sometimes, Ines explained that she felt as if all her efforts had amounted to nothing but a sense of feeling like a 'foreigner everywhere'. In spite of acquiring citizenship and all the years working in the hospitality industry, Ines remained an 'other' in Switzerland and in the valley: locals still described her as 'the Portuguese saleswoman'. But with the majority of her existence spent in the valley, she also no longer felt like she fully belonged to her native Portugal. Coming 'from here and there' (*ici et là-bas*) was an opportunity – she said she had learnt a lot – but it also meant being 'from nowhere' (*nulle part*).

Gaining a deeper understanding of the struggles of workers like Ines as my fieldwork progressed, the tourism expert description of the relationship between the resort and migrant labour as a 'win–win' situation seemed increasingly at odds with their vision of the situation. But his description entailed temporal claims that drew my attention. In his words, work and wages marked migrant stays in Switzerland, whereas life-building was an activity to be led in the latter's home countries, when they go back to make a future 'of their own' with the money earned in the wealthy Swiss economy. This reflected a certain truth, as evidenced by Ines's desire to return, a sentiment shared by many of her compatriots.

As various authors have shown, Portuguese migration, particularly for the first- and second-generation migrants, has been return-oriented (Klimt 1989, Graf 2021). The simultaneous involvements of these migrants in their home country of Portugal and in their countries of residence speak of a transnational reality that extends beyond the 'containers' of the nation-state (Glick Schiller 2009). They also reveal a temporal orientation marked by a present of temporariness lived abroad and a future of permanence to be lived in Portugal (Klimt 1989). However, general narratives often tend to mask the diverse perspectives on return and the desire to remain in destination countries – a phenomenon highlighted by Caroline Brettell in her comparative studies on the Portuguese diaspora (Brettell 2003; see also Graf 2021). At the same time, she notes, 'attitudes about returning to Portugal are affected by the attitudes of the host society toward immigrant groups, by work laws, by the susceptibility to exploitation, and by the fear of expulsion' (Brettell 2003: 194).

The context of the Alpine village, Switzerland and Western Europe at large, reflects such reality, not merely reacting to but actively shaping the temporariness of migration – whether from Portugal or elsewhere. Although the presence of migrant workers had been recorded from the end of the nineteenth century in the valley, my interlocutors, sometimes migrants themselves, repeatedly spoke of foreigners in such temporary terms. When they even evoked them, many village inhabitants called migrants who were active in the hospitality industry (especially in the lower positions) 'guest-workers' (*Gastarbeiter*) or 'seasonal workers' (*Saisonniers*), sometimes even though they knew they were permanent residents. The reasons for such an exclusion in the context of touristic success and historical longevity bring us back once more to the Swiss history of *Überfremdung* (see Chapter 4).

THE MAKING OF FOREIGNERS IN SWITZERLAND

From the mid-nineteenth century, Switzerland ceased to be primarily a country of emigration and became one of immigration (Vuilleumier 2012). In the 1880s, migrants, mostly from Italy, were recruited for the realization of large infrastructural projects such as the construction of tunnels and roads (Piguet 2013). In the Alps, the deployment of an impressive touristic infrastructure was emblematic of this process, with a strong reliance on workers from Italy and, to a lesser extent, Germany (see Moser 1997). It is in this context that the central idea of *Überfremdung* – 'foreign infiltration', 'over-foreignization' or 'alienation'[5] – gained importance in Switzerland. Within this climate, foreigners cast a threat on the integrity of the Swiss national community (Kury 2003; Dahinden 2016). With the foreign infiltration thesis, migration transitioned from a municipal or cantonal matter into a national concern on who 'ought to' belong to the country, marking an important step in the 'ethnicization' of belonging to the Swiss community (Wicker 2003). At the beginning of the twentieth century, these concerns reached the highest political level of the country, the Federal Council, which adopted a more restrictive attitude towards foreigners with the deployment of an Alien Police (*Fremdenpolizei*) and the creation of a national Alien Act (*Bundesgesetz über Aufenthalt und Niederlassung der Ausländer*). These institutions defined the contours of an 'acceptable' migration under the following statement, communicated in 1924: 'There will be . . . nothing to object to the influx of foreigners, only provided that the latter do not think of settling in' (cited in Piguet and Mahning 2000: 4).[6]

In 1931, the Alien Act came into force, and with it the creation of the *Saisonnier* (seasonal or guest-worker) status that mediated between national economic interests requesting 'flexibility' and the risks of the cultural alienation of Switzerland (Arlettaz 2012). When Switzerland entered a favourable phase of economic growth after the Second World War, the *Saisonnier* status regulated the flow of migrants, from Italy in particular, whose stays were limited to nine months (Piguet and Mahning 2000: 4). Half of the foreign workforce present in the country was seasonal at the time (Fischer and Dahinden 2016: 10).[7] But, at the heart of the authorities' preoccupations was the worrying possibility of a bad economic conjuncture, potentially leading migrant workers to 'burden' the country with high social security costs (Niederberger 2004: 48). Such precautionary measures led to what sociologist Josef Niederberger has described as an explicit 'politics of non-integration' (ibid.: 41), based on the temporal and spatial segregation of all migrants, but targeting seasonal migrants in particular. Limited to temporary permits, migrants also inhabited a different space from their Swiss counterparts, often in barracks and/or in close proximity to their workplace (Fischer and Dahinden 2016: 10). Family reunification policies became particularly restrictive, to counter migrants putting down 'roots' in Switzerland, and to facilitate potential deportation operations (Niederberger 2004: 48).

In the 1960s and 1970s, the narrative of the *Überfremdung* took a more openly xenophobic turn as it revolved around the cultural consequences of immigration on native Swiss values (Wicker 2003). Inscribed under Article 16 of the Alien Act, *Überfremdung* became a legal criterion, decisive for the fate of foreign applicants with regards to their economic and cultural impacts on the Swiss nation. At the same time, the so-called Schwarzenbach initiative (1968), in particular, aimed to counter the foreign alienation of the country. Named after its founder, the far-right politician James Schwarzenbach, head of the National Action Party, the text demonstrated xenophobic, racist sentiments towards foreigners, particularly Italians. Schwarzenbach proposed to limit the presence of immigrants to 10 per cent of the Swiss population and to deport 350,000 foreign workers to reach this end (Landwehr 2020). The Schwarzenbach initiative, put to vote in 1970, was rejected by 54 per cent of Swiss voters. However, in the years that followed, the Swiss state introduced foreigner quotas as a coping measure against the oil crisis hitting Switzerland (Piguet and Mahning 2000). Many of the Italian workers left the country, and were soon successively replaced by Spanish, Yugoslavian and Portuguese nationals, active in diverse sectors such as the construction, agriculture or hospitality industry.

Even when the Alien Act underwent full revision and allowed for a greater differentiation between the profiles of migrants sought by Switzerland in the 1990s, the *Überfremdung* narrative remained central, although repurposed to respond to the new globalized conjuncture (Piguet 2013). Unqualified migrants attracted by the guest-worker regimes were then seen as obstacles to the flexibility and modernization of the Swiss labour market (Afonso 2006: 110). After refusing to join the European Union, Switzerland also sought to normalize its relation with Europe regarding migration and other matters of exchange. Riaño and Wastl-Walter describe the Swiss situation at the time as an ethical quandary, with Switzerland attempting to 'move close to the EU despite fears of *Überfremdung* among the Swiss population' (Riaño and Wastl-Walter 2006: 1702). In their view, core concerns over foreign alienation were addressed with the design of a three- and then two-circle policy, conflating spatial and cultural distance.[8] The Swiss regulations expected nationals from the inner circle to integrate more seamlessly in Switzerland (ibid.).

More recently, Switzerland has redefined the profile of migrants it wished to attract by recruiting a highly qualified workforce from the so-called 'third countries', in response to a lack of workforce in the country's top industries, such as pharmaceutics and finance (Sandoz 2019). Class and culture have increasingly served as a marker of integration and an easing of cultural proximity. Reflecting these policies, migrants in Switzerland (approximately 26 per cent as of 2021) have occupied the higher and lower ends of the job market,[9] while the *Saisonnier* status was officially abolished in 2002 with a bilateral agreement on freedom of movement. Yet, far from disappearing, the *Überfremdung* discourse has continued to shape the changing criteria of migrants' desirability and their alienating character.

Ines, like most of the long-term migrant residents I met in the village, remembered the time of her arrival, the 1980s, as difficult because of the consequences of policies based on the fear of foreign alienation. Under permit A (the *Saisonnier* permit), she was, like a great number of her siblings, active in the hospitality industry, with hard, long and scattered working hours, restricted free time, and the low wages associated with typical guest-worker jobs, which were unattractive and relatively easily avoidable for many Swiss – all typical features of the hospitality jobs in the country, the Alps, and indeed across the world (Jordhus-Lier and Underthun 2014; Ioannides and Zampoukos 2018; Graf 2021). Ines was repeatedly forced to leave the valley, taking all her belongings with her at the end of the season, not knowing whether her contract (and thus her permit) would be renewed the following year.

Ines remembered how the Italians she worked with were still insulted and mistreated: many Schwarzenbach supporters, but also Swiss people more broadly, continued to draw on the cultural, ethnic and racial incompatibilities between Swiss natives and Italian immigrants formulated in the 1960s. Italians at the time were portrayed as threatening, dark-skinned, sexualized and careless 'others' (Maiolino 2012; Jain 2020). This story of discrimination was vividly remembered by Ines but also by many other migrants, Italian or not, who remembered hearing about Switzerland as the place where some restaurants were 'forbidden to dogs and Italians' (see Maiolino 2020). Clara, an Italian worker, told me about her father's concerns when she chose to come to Switzerland in the 1990s, warning her about its xenophobia. After the Italians left, the Portuguese replaced them as the new unwanted, Ines said, treated like 'donkeys' or 'slaves', and rejected by the Swiss.

Ines's description echoed some of my discussions with Swiss locals. If tourism had pulled the village out of poverty, as the modernist narrative had it, the Swiss and *Einheimische* locals I met showed a strong reluctance to work in the hospitality sector, and not solely because of the poor working conditions associated with it. I was told multiple times – by locals ranging from a hotelier and a local politician, to a retired native woman – that service was not made for the Swiss as 'the Swiss was not so much of a servant' (*der Schweizer ist nicht so der Diener*). Given the nation's culture and its ideals of freedom and democracy, shaped by the Alps as the hotelier reminded me, the Swiss and *a fortiori* the native locals 'as *Eidgenosse*' culturally did not fit in with the service economy (see also Chapter 1). According to him and other locals, nationals of other countries who could not draw on such roots and who had historically all lived under 'imperial rule' were less bound to ideals of democracy and thus logically fitter to serve. Another of my native local interlocutors claimed that the allegedly 'hot, Mediterranean-type climate' from which the Portuguese came made them ideal workers for the hotels' steamy laundry rooms. Claims on history, the environment and corresponding local and national cultures were often interwoven with the economic rules of the liberal division of labour – a common association in light of the history of the Swiss *Überfremdung*. Together, these discourses shaped a highly stratified or segmented local economy, not only between the tourists and the toured but also *among* the latter, between Swiss and/or native locals and migrant intermediaries, mirroring what has been described for other resorts and within hospitality businesses (van den Berghe 1992; Adler and Adler 2004; McDowell, Batnitzky and Dyer 2007).

OTHERS FOREVER

The 'business of foreigners' and nativist sensibilities have more in common than their contradictory features suggest, for instance when they serve to shape a hierarchy of guests according to their cultural and historical proximity. The question of migrant labour in tourism – an undeniable global reality (Zampoukos 2018; Janta and Ladkin 2024) – reveals yet another involvement of this industry, although imbued in progressive and cosmopolitan imaginaries, with exclusionary politics. Like the construction, agriculture and care sectors, the hospitality business has historically relied on the labour of populations marked by their difference, be it gendered, racial, ethnic or national, to accumulate capital (Holmes 2013; Kasmir and Carbonella 2014; Ladkin 2014). The latter represent a cheap and flexible workforce in such particularly labour intensive and fluctuating sectors (Janta et al. 2011).[10] Marxist scholars of migration in particular have described how migration regimes have shaped the conditions of importation and exploitation of the workforce, while exporting the costs related to unemployment, health care and education to the workers' countries of origin. The exploitation of the latter is separated 'from the sites (and costs) of [the workforce's] reproduction' (De Genova 2019: 435; see also Burawoy 1976). Spatial imaginaries of difference allow for and justify such forms of migrant exploitation, visible in the Swiss three- or two-circle policies or the emergence of a Schengen space discriminating between EUropean nationals and the rest.

Regardless of their individual prospects for return, the experiences of people like Ines – shaped by the temporary guest-worker regime – and the enduring local narratives framing hospitality workers as temporary, despite their permanent presence since the nineteenth century, underscore the inherently temporal nature of migration. Anthropologist Ayşe Çağlar highlights this aspect when she speaks of particular 'chronotopes', which structure scholarly, policy and everyday discourses on migrant presence within given national communities (Çağlar 2018). The prominent integration chronotope[11] 'situat[es] migrants in space and time in a particular way, namely in a linear and a sequential process', and requests of the latter that they 'integrate themselves into the countries of settlement *in time*' (Çağlar 2018: 27, italics in original). Based on Johannes Fabian's analysis of the 'temporal othering' of research subjects in anthropology (Fabian 1983), Çağlar posits that a 'non-synchronism' (Çağlar 2018: 34) of migrants guides integration policies inscribed in legal texts, public discourses and everyday representations of foreigners in many global contexts.

In Switzerland, the particularly restrictive immigration policies and notoriously strict citizenship rules[12] subordinate a foreign population – a quarter of the country's inhabitants – to an acute form of the integration chronotope. Migrants are judged against conditions that are proof of good integration, such as their mastery of national language, their integration into Swiss social networks, their civil obedience, and their involvement in the economic life of the country when renewing their permits and/or applying for citizenship.[13] These measures point to integration as a liminal, 'perpetual state of becoming' (Çağlar 2018: 27) in which migrants find themselves de facto under the umbrella of the Federal Act on Foreign Nationals and Integration. Under the *jus sanguinis* rule, which grants citizenship based on 'blood ties' or progeny, this process of perpetual becoming stretches beyond generations (Bauböck 1998). Migrant presence is essentially framed as temporary, revocable and conditioned to the denizen's capacity to integrate into the normative, eternal space *and* time of the 'host' society.

Ines in many ways embodied an exceptional case of successful integration: she spoke the language, worked among Swiss, and had even obtained citizenship, unlike many of her compatriots. She had reached the ideal future of full integration that migration policies foresaw for foreigners like her. Yet, she also described a persistent sense of feeling like an 'other', coming from another time and space – that of Portugal – wishing but also having to leave soon for another time and space upon retirement because of her inadequate retirement savings, eaten away by years of small salaries and no contributions under the permit A. The integration chronotope still shaped Ines's world, a world of 'injustice', made up of an impossible gap to be bridged between 'us' and 'them', she said. A gap that her children, some of whom had also acquired citizenship, experienced in the same way, she told me. 'We are not like you, not like the Swiss', she once responded when I enquired why her children had left the valley where they had been born. The present, comprising the ongoing consequences of an economic, legal and social system, had for decades been treating her, and her children, as non-integrable, temporary but exploitable 'others'. Leaving the valley seemed like an obvious choice for them, in contrast to those who had been made to feel like they belonged.

In Switzerland, many have portrayed the guest-worker regime as a closed chapter of the country's history: one ending with the abolition of the permit A in 2002, following the bilateral agreements between Switzerland and the European Union on the free movement of individuals (AFMP) (Riaño and Wastl-Walter 2006: 1703). Journalists and academics have recently uncovered the untold stories of exploitation, racism and precarity experienced by many *Saisonniers*, many of whom were Italian, to reveal the harsh treatment of these workers in Switzerland in the Schwarzenbach years (Jain 2020; Falk 2022). Following this, accounts of contemporary migrant precarity in Switzerland now focus on the brutal experiences of refugees, asylum seekers and illegalized migrants, who present similarities with the historical *Saisonniers* as a temporary, deportable workforce (Amarelle 2010). Jacqueline Kalbermatter has for instance demonstrated how refugees with insecure residence status have been integrated into the hospitality industry, which depends on a low-cost workforce (Kalbermatter 2020). By contrast, European migrants are said to enjoy rights comparable to the Swiss, with the exception of voting rights (Riaño and Wastl-Walter 2006).

EU nationals undeniably benefit from freedom of movement and they experience radically different realities of migration from third-country nationals, who are subjected to greater forms of discrimination. For refugees like the Afghan boys I met in the valley, a sense of profound exclusion and uncertainty shaped their existences from very early on, not least because of political discourses insisting on their cultural incompatibility posing a threat to the Swiss nation (Lems 2022). But, years after the formal abolishment of the guest-worker regime linked to permit A, my findings point to the ongoing effects of the integration chronotope on all migrants in Switzerland, Europeans included. This appears especially important with the recent waves of toughening integration measures and the subjection of migrants, as well as their children and grandchildren, to Switzerland's conditional hospitality such as the revision of the 'Federal Act on Foreign Nationals *and Integration*' in 2019, officializing more restrictive measures towards foreigners under the guise of 'integration'.[14] In spite of being born on Swiss territory, their presence is essentially framed as temporary and revocable if the criteria of cultural and economic integration are not fulfilled. Even when these criteria are formally met, as in Ines's case, experiences of temporal 'othering' stretch through one's existence and beyond. In Çağlar's words, this situation makes for a 'denial of shared temporality', discriminating between two ontologically different trajectories, two different worlds: the migrant and the non-migrant world (Çağlar 2018: 29; also 2013).

WORLDS APART

Many migrant workers I met in hotels and restaurants of the valley accounted for a feeling of inhabiting a different chronotope, making up a foreigners' world or *Ausländerwelt*, shaped by 'foreigner' problems and experiences, which were different from those of the Swiss and *Einheimische*. This world was one of small apartments, often adjacent to the village's main street or

Hauptstrasse, where the flags of home countries sometimes decorated walls, windows and porches. It was one of direct proximity with their workplace, as migrants often worked irregular hours, often two shifts a day. These spaces were sometimes subjected to the state's surveillance, like for Andrea, a hotel worker who had recently arrived from Portugal when I met her. She struggled to find any accommodation close to work and school at an affordable price in a touristic valley where rent was skyrocketing and where her employer had no employee rooms available. When she did find an apartment, it was inspected by the local immigration authorities to guarantee that its size was appropriate for family reunification, which added to her stress.[15] 'It's a foreigner home (*Ausländerwohnung*), it's small', she told me, unsure about what proportions were deemed suitable for the authorities and whether she would receive a letter from the Alien Police rejecting her application or requesting that she find a bigger place on her small salary.

For the Italian hotel worker Clara, who had lived in the village for the past fifteen years and was a beneficiary of the secure C permit (establishment permit), living in a foreigner world meant reasoning with the possibility of losing her job when her already aged boss would retire. Clara referred to the example of another hotel where a new owner had decided to change its entire personnel. She told me that if that were to happen to her, she could become jobless and lose her and her family's right to stay in Switzerland. Her level of German was still poor, compared to her Portuguese, which proved much more useful at work. She had no education in the hospitality industry, a bad back, and was soon to turn fifty. 'Who would employ me then?', she asked me once with a worried look on her face. While the C permit had granted her security until then, she feared being stripped of her right to stay because of her lack of 'integration' if her employment situation were to change. Her Spanish partner, although he had worked in the village for the past seventeen years, was no more secure either, as his job was still seasonal and so his contract renewed every year. Both found themselves with an insufficient level of oral and written standard German to pass the citizenship test, and without the money or time to prepare for the long submission process.

These experiences point to the consequences of the temporariness or 'temporal othering' induced by the integration chronotope, and I will describe how migrant workers actively dealt with and responded to this uncertain situation in the next chapter. Yet, as is implied in Mikhail Bakthin's definition of the 'chronotope' (literally meaning 'time-space') on which Çağlar relies (2018), additional layers of spatial and, more broadly, social concerns, 'fuse' with the temporal into a 'concrete whole' (Bakhtin 1981: 84).[16] These spatio-temporal layers structured the ways in which migrants and Swiss locals occupied different positions within the village.

One place that undoubtedly stood out as a foreigners' world were the 'backstage areas' of hotels and restaurants. While I conducted participant observation in one of the resort's hotels, one of the housekeepers from Croatia noted, 'It's always been like that, always foreigners', as we were talking about the personnel. The team of chambermaids walking through the hallway with us came from Portugal for the most part, and had replaced Italians and Spaniards. A few of them were now part of the 'new trends' of people coming to the village and Switzerland from the East, such as Slovakia, the housekeeper explained. The 'business of foreigners', the housekeeper explained, had historically evolved into a business based on the welcoming of foreigners (tourists) by other foreigners (migrants). What had been framed as an exceptional, short-term policy measure to address labour needs by recruiting a foreign workforce through the *Saisonnier* regime more than a century ago, had become a long-standing feature of the local and national economy – a pattern described by other contemporary analysists of temporary migration programmes in Canada (Foster 2012; Bryan 2018).

During my participant observation in the hotel, Germans, Portuguese, Polish and Slovakians all explained how this segmented economy translated into them having very little contact with the Swiss- and/or *Einheimische* villagers. The foreigners' world opposed the 'bubble' that Ines regularly spoke of when she described how Swiss and native local villagers lived. Particularly irksome to her was that they chose to open this bubble up whenever they wished to enquire about her life, but quickly closed it off again when she wished to talk to them. The boundary abruptly reappeared even as Ines thought she had made friends with other mothers waiting for their children in front of the school, for instance. 'They just closed the door, and it hurts', she said, recalling some burgeoning relationships that had quickly fallen apart for no apparent reason. But she did not always incriminate the Swiss, broadly speaking: this attitude was also a localist one. 'Even those from the next valley are not really theirs', Ines said, referring to the lines of (un)belonging I described in Chapter 1. However, she knew it was different in her case: whereas it was possible, though difficult, to be more or less accepted in local networks as an outsider – especially an outsider woman from another Swiss-German region – this was not the case for outsiders like her.[17]

Many Swiss and native locals, too, acknowledged the bubble they lived in. On the one hand, they stated that it resulted from a cautious and cold attitude towards outsiders determined by their natural home, the Alpine environment and/or an economy of fleeting stays of both tourists and migrant inhabitants. On the other hand, they considered migrants, particularly Portuguese, responsible for their reluctance to integrate, for staying among themselves, for speaking their own language – for living in their own world. The Portuguese were a 'people of their own' (*ein Volk für sich*), as one villager once put it. The gap separating the migrants and non-migrants in the valley was manifest in the absence of stories of intimate relationships and friendships between migrant hospitality workers and Swiss-born villagers.[18] At the same time, this internal boundary also impacted children, as Clara's son, aged ten, for instance told his mother when he came home from school. Born and raised in the valley, he was being bullied by Swiss kids, especially of native families, who mocked him for not having a Swiss passport. Ines's children, too, had been harassed in the local school because of their non-Swiss origins.[19] One retired teacher corroborated these experiences when she explained to me that the divides in her classroom between Swiss native villagers and the others were tangible, and reflected the exclusionary atmosphere of the valley.

The experiences of migrant workers like Ines, Clara and Andrea shed light on a certain continuity of the *Saisonnier* regime under the integration chronotope. They reveal the temporariness with which migrants remain considered in Switzerland, following the abolishment of the guest-worker status. One additional and perhaps most tangible trace of the persistence of the *Saisonnier* regime is the fact that, in the valley, a third of the local foreigner population remained subject to temporary L permits in 2019, according to the municipality's records.[20] The latter status granted foreigners the right to remain in the country during their employment in the touristic high season in hotels and restaurants, for durations of less than a year. Most of the hospitality workers I met saw the short duration of the L permit as a disadvantage, especially those who had families in the valley. Although bound to more rights and social benefits than the abolished *Saisonnier* permit,[21] the L permit had to be renewed every year.

This made for difficult situations, akin to what has been discussed by Catherine Bryan in her work on temporary migrant hospitality workers in Manitoba (Bryan 2018): uncertainty regarding the next season, employment relations and family reunification was permanent (see also Adler and Adler 2004; Binford 2013).[22] Andrea, the Portuguese worker mentioned earlier, had for instance been working and living in the village under the L permit regime for three years as a waitress in a hotel when we met. She was desperately seeking a permanent position so she could finally obtain a B permit (resident foreign national) for herself and her family. When she finally found a permanent contract, she called it a 'luxury', a precious and rare gem in the seasonally led business of tourism. Andrea's friend and his wife did not have this luck and so were both still living under an L permit after three years in the village, and now that they were expecting a child they felt particularly concerned about the precarity that was bound up with this status.

TOURISM DEPENDENCY BETWEEN HEIGHTENED AWARENESS AND INDIFFERENCE

Along my research, I repeatedly encountered ambiguous evocations of the common good in a divided, conflict-ridden context. The *Touristikers'* promises of a common future of prosperity, for instance, clashed with the visions of those in charge of the valley's commons: the Alpine corporations. Yet, in spite of their disagreements, my (Swiss) interlocutors shared a similar vocabulary, evoking the idea of being affected in economic, material, social and emotional ways by tourism, as its dependents. They associated tourism's (mis-)developments, its slipperiness, with the threatening possibility of losing their means of subsistence, their land, their friends, their families, and the chance to secure their heritage and their future. Both opponents and supporters of the 'project of the century' were united by the belief that if the wrong decision were to be made regarding the touristic development of the region, this could have tragic consequences for a business that had created everything the present offered in the valley. The play *Alpine Glow* presented a similar take on the excess of innovation.

At the heart of these local narratives and experiences was the possibility of being dispossessed of everything that the present offered, to become nothing again, by going back in time to face the challenges that their ancestors had faced before tourism, or by moving forward endlessly into a future of increasing alienation. Ines's apparent disinterest in events that I had come to understand as manifestations of tourism dependency in the village during my encounters with Swiss and/or native locals thus struck a particular chord. In spite of her apparent lifelong and strong dependency on tourism as a source of employment, allowing her to migrate and stay put in the valley with her sons, she seemed detached from its developments. Where tourism dependency, as an unstable business based on the wanderlust of others and a changing global constellation, was characterized by a form of heightened temporal awareness for the native villagers, it seemed to induce a form of indifference for Ines and many of her migrant peers. Who, then, were those implied in the narratives of the common good, the community, the valley that stood at the centre of the new cable car and the historical play?

Following Marres' approach revealing the power of projects to 'make' publics (Marres 2012) instead of simply drawing on pre-existing audiences, the remainder of this chapter addresses the exclusionary dimension of the temporal projects I have examined in the previous chapters. At its core lies the uneven distribution of the capacity of locals (migrants and non-migrants) to identify with and feel affected by the valley's tourism development.

NATIVE SUBJECTS AS AFFECTED DEPENDENTS

After the voting 'of the century', the JR company and the *Touristikers* put forward an idea that their project had become an object of the common good, legitimized by the processes and values of democracy. The 'impacted', concerned population, as tourism dependents, voiced their opinions, and their right to self-determination had to be respected, the company and its supporters argued. Yet, looking at the voting more closely, this exercise of democracy (like it often is) appears to be more exclusive than it was purported to be. Out of the almost four thousand people living in the village, only 1,466 locals utilized their right to vote. While this represents a minority, the voting was in fact very well attended in comparison to national attendance rates (between 40 and 50 per cent in recent years), with 72 per cent voting of the approximately 2,600 people entitled to vote.

Voting rights were defined under the following conditions: being a Swiss national, aged eighteen or above, and resident in the village for at least three months. These conditions, set out in the national and/or cantonal regulations, delimited who could exert their political rights to ensure democratic representativeness. By restricting rights to vote to nationals, these regulations nonetheless follow an exclusionary idea of democracy that has not always been a given in the Swiss context.[23] According to Rosita Fibbi, the translation of Swiss nationality into citizenship is the result of a political project of ethnic and cultural redefinition of political rights initiated at the end of the nineteenth century (Fibbi 2011: 8–9). Acknowledging the discriminatory nature of the exclusion of foreign inhabitants, some cantons and Swiss municipalities located in more liberal areas than the Bernese valley have allocated migrants the right to vote on communal matters.[24]

On one of my visits to Franziska, she complemented such legal definition of democratic will by telling me that every legal voter's opinion on the valley was not equally valid or respected: a version of *Einheimischness* was essential when it came to defining whose voices and opinions on tourism development were deemed hearable. Referring to a man who had settled in the valley after retirement and who wrote a reader's letter in the local newspaper to denounce the problems regarding one of the many hotel ruins, Franziska and her husband contended that his opinion did not really matter. As an urban dweller who had previously 'only' owned a holiday home in the village, who did not make a living here and who had 'just' come for retirement, this man could not have any great understanding, or deep knowledge, of what living in the village meant. Franziska turned to me and said: 'As one of my sons likes to say, "One can become Swiss but one will never become a confederate (*Eidgenoss*); one can become an *Einheimische* but you must have a *Vorsass*"'. I quickly scribbled down this sentence, which Franziska helped me to unpack over the following months, going through the multiple layers it contained.

With this explanation, Franziska revealed a nativist understanding of dependency: it defined who ought to genuinely feel and be seen as dependent beyond criteria defined by the already exclusionary yet democratic process of voting. Speaking of the *Eidgenosse* evoked a nativist narrative according to which migrants would always be foreigners; even when in possession of the citizenship they would not be real, natural, original Swiss – an idea defended by the far-right SVP in Switzerland (Skenderovic 2007), for whom Franziska's sons voted. At the same time, although the Swiss lowlander and former holiday-house owner 'was allowed to give his opinion' on votes like all Swiss citizens living in the municipality, Franziska repeated her son's saying because she believed that his voice was less legitimate, in a more social sense, than those of the real dwellers – the *Einheimische* like her. Owning a *Vorsass*, the traditional local Alpine hut, was associated with this status. At first, becoming a native via the acquisition of a *Vorsass* seemed to reveal a more flexible definition of belonging than that of the ethnic one evoked in the impossible becoming of an *Eidgenosse*, or authentic Swiss, for foreigners. In practice, however, Franziska herself stated that acquiring a *Vorsass* was a very complicated endeavour that even money could not buy because these were goods circulated among native locals. As already shown (in Chapter 1), other criteria such as speaking the local dialect and carrying a typical family name were of great importance in order to be viewed as a genuine native.

Franziska's explanations point to the idea of being dependent on tourism as something inscribed in one's own history, one's own subsistence and survival in relation to tourism – something that was common to many *Einheimische* I met. The latter believed that no one else besides them, as traditional *Bergleute* (mountain people) living off tourism in the Alps and building on a particular history, could understand what it meant to rely entirely on tourism as an economic activity, and the dilemmas this brought. Urban lowlanders who came to settle or buy a holiday home in the village could not grasp what it meant to have one's subsistence depend on tourism in an Alpine area, as I heard countless times from mostly *Einheimische* (for similar cases in the Alps, see Petite 2013). Like one of Franziska's neighbours from the lowlands, these inhabitants showed a certain disdain vis-à-vis the locals, lecturing them on how to live in the mountains when genuine people like them already knew in their bones and history what it entailed. Only the *Einheimische*, Franziska and her peers retorted, as the affected people, in their histories and experiences, could truly make legitimate claims on tourism dependency.

The theatre play *Alpine Glow* underlined such a subject of dependence too, by casting the *Einheimische*, and solely them, as *the* affected population – impacted, benefiting and/or losing from tourism – materially, symbolically and emotionally. To take up on the play's wording, the local natives represented the eternal hosts, who genuinely and naturally 'belong(ed) to the valley and the valley belong(ed) to them'. The (European) guests were incorporated into a 'community of loss' together with the natives (Berliner 2012), both in the story told by the play and among the audience it was targeting (see Chapter 3). Yet, the play clearly and dramatically identified the losers of this tale: the (native) locals and their youth, who would not 'inherit' anything and who would be forced to outmigrate, like their village spirit.

In the valley, the centrality of the *Einheimische* as *the* affected dependent often appeared in its potential status as 'exiled' victim of tourism's misdevelopments. While most locals acknowledged the positive migration balance that the valley could draw on, as one of the few mountain communities to gain new inhabitants, they also repeatedly evoked outmigration as a menacing, omnipresent possibility. Many villagers, often natives and sometimes in important political positions, worryingly told me about young native families who had migrated towards the lowlands because they were unable to find affordable housing in the village on account of the increasing gentrification and regulation of the region.[25] However, my attempts to find figures to support these claims were unsuccessful.[26] In spite of the difficulty in attesting to the actual reality of this phenomenon, these narratives revolved around the natives' exile as a particularly abject scenario: that of uprooting the naturally rooted ones – a motive regularly evoked in Romanticism (Starobinski 1966). At the same time, various villagers shared concerns that another type of immigration – of foreigners – was affecting the 'core' of the village. This did not apply to all of my respondents; Franziska noted the comparatively greater influence of tourists over migrants at the beginning of this chapter. However, some natives believed that migrants could be responsible for the eventual disappearance of the next generation of genuine locals in the region (see Chapter 1, and Messerli 1987).

INDIFFERENCE AND MUTED DEPENDENCY

From a boundary-making perspective, what unites some around an idea will, by definition, exclude others (Lamont and Molnár 2002). The making of the *Einheimische* as the epitome of the population of the affected natives, the eternal inhabitants and hosts of the valley around the idea of a common teleology of touristification, and the fear of dislocation, reiterated by the cable car project and the play, therefore de facto excluded others. Outsiders, like the lowlander neighbour of Franziska, but also and especially foreigners, were assimilated into the world of the 'travelling ones', not the 'toured ones'. Such boundaries had the effect of overshadowing the role of people like Ines, Clara and many other migrant hospitality workers, who were indeed part of both the travelling and the travelled populations. Migrant dependency on tourism, contrary to the situation of retired urbanites, was at least as great as that of the natives, with economic integration representing the basis for their and their families' right to stay, as Ines's life story demonstrates. How then, could such a sense of indifference emerge from their side?

In some cases, migrant hospitality workers – mostly the newly arrived ones, who had been in the village for less than two years – did not know about the cableway project and learnt about it through me. In contrast to many Swiss and *Einheimische* locals, who were more or less involuntarily involved in networks of information from the beginning of the project, these workers were excluded from what seemed to be common knowledge and concerns. As foreigners, their rights to vote were denied for the cable car project, like for any other voting, whether local or national. For many of them, their presence was restricted to seasonal or yearly permits, and their contacts with the local 'deciding' population limited. They often also lacked a sufficient level of Swiss-German and/or German to follow debates in the newspapers or in assemblies. Clara, who paid for a local newspaper subscription so she could stay informed about the valley's events, admitted not being able to read it. The papers piled up on her kitchen table before she threw them away, often without having had the chance to look at them, she told me, visibly embarrassed. Information on the debates that were said to divide the valley did not reach many of these migrant workers, other than coincidentally, for instance through me.

When I asked migrant hospitality workers who knew about the cable car what they thought about it, they stated that they themselves had no view on it, or if they did, they kept quiet about it. Andrea, the newly arrived Portuguese woman I introduced earlier, explained after she had learnt about the project through me that she 'did not have any opinion' because she was 'here just to work'. A bad word on her part could cost her a lot, as happened to a colleague of hers in a previous season, who lost her job for gossiping. The same sense of caution was adopted by an Austrian train mechanic, who explained to a group of tourists when they enquired about his opinion on the new cable car that he 'did not wish to give his opinion as a foreigner'. Although the presence or absence of the infrastructure would most likely deeply impact his everyday activities, his statement, like Andrea's, revealed the strong sense of risk involved in positioning oneself for or against the cable car for Swiss or native locals, but also for migrants.

For many hospitality workers, such precautions were also a matter of integration into a hostile environment. Dušan, a hospitality worker from Serbia, contended that his efforts to be polite, sociable and likeable were essential to overcome the label of 'shitty foreigner' (*scheiss Ausländer*), with which he was assigned from the beginning. 'If you do something wrong, they will remember', he said. 'You have to show patience as a foreigner.' Every day, Dušan insisted on greeting locals who did not respond, explaining that 'eventually they will say hi back'. Publicly risking an opinion on the contentious topic of tourism development in this context seemed to be particularly risky for migrants. 'We're guests (*Gäste*) here, we have to adapt', Dušan concluded. When I commented that, after twenty years of residence in the valley, he should no longer be considered a guest, he responded with a cynical look that felt like an answer to the naivety of my statement; he then changed the topic.

Ines's case, although echoing some of these stories, was different: in contrast to most migrants I have talked about in this chapter, she had acquired citizenship and the right to voice her opinion shortly before the voting on the project of the century, but made the choice to not use her vote. She was generally sceptical towards democracy in both Switzerland and Portugal, which she believed was based on corrupt politicians and the favouring of the rich. 'The only difference is that here [in Switzerland] it is undercover (*caché*) and there [Portugal] it is open.' But crucially, Ines also did not vote because of her deep-seated feeling of unbelonging, along with the consequences of decades of being made to feel different. Her experiences of being casually excluded from local networks, especially the one where other parents got together when their children were in school, left a mark: 'I felt like I was nothing, they did not talk to me anymore, they made me into less than nothing (*une moins que rien*)'. After learning that even her children, born and raised in the valley, were experiencing instances of bullying at school, our discussions often revolved around the multiple ways in which she, like other Portuguese, had been made to feel foreign. 'Why would I vote then?', she asked me once, before stating: '[This village] is not my village! In a way it is, but in another it isn't, because I don't feel Swiss, Swiss a hundred per cent. Even if you came to live here, you are Swiss, it's like you are part of this world – but in reality, you aren't because you are not from [the village]. You [as a Swiss] would only be accepted in the village if you marry one of them'.

Ines's statement resonates with Franziska's quote on the *Eidgenosse* and the *Vorsass*, where she referred to the nativist politics of belonging, between localism and nationalism. As a migrant, Ines knew that she would 'never' become a Swiss, in the cultural or ethnic sense of the term, especially in an area where even other Swiss people did not succeed in seamlessly integrating into the community. She also believed that there was no possibility of her 'making money' around the cable car in the way promised to the *Einheimische*, who had property in the village, good jobs, and opportunities tied to JR. By contrast, her own horizon was defined by financial struggles, which had started from her teenage years, when she was forced to work in the fields in Portugal with her parents. These struggles made her leave Portugal for Switzerland at the age of seventeen, and they were now playing into her decision to leave again, for Portugal this time, where she preferred to finish her days. In five years' time, she knew that her retirement pension would not allow her to survive in Switzerland, and she already felt like she was 'just working to pay bills'. A life made up of years without social benefits under the permit A, and then bad salaries, seasonal contracts (with months of unemployment in between) and a costly divorce had all made her 'poor', Ines said.

Language and social barriers, the strong segmentation of the labour market, the pressure associated with integration, and enduring feelings of humiliation, all factored in the creation of the unawareness or indifference of migrant workers to communal, future-oriented forms of temporal reasoning, like the cable car. The theatre play *Alpine Glow* constituted another moment of exclusionary temporal reasoning that further exemplifies the non-recognition of foreign hospitality workers as community members in more historical terms. While some of the latter explained that they did not know about the play (the advertisements in the newspaper were in German), others explained they had 'no time' to go to the play in the middle of the high season. Ines invoked financial reasons besides the lack of time; sixty Swiss francs was simply too expensive for her.

Absent from the audience, migrant hospitality workers were also absent from the content of the play, with the entirety of characters representing either Western tourists or native hosts. During my interview with one of the theatre organizers, I asked him what happened with the many outsiders, such as foreign hospitality workers, who had also helped to make the resort what it was today, in the 'historically accurate' portrayal of the past. The organizer remained silent for some time before explaining that the play was an artistic product and not a 'social survey', because they 'only had one hundred minutes after all' to work with. Other locals reacted similarly when I questioned them on the absence of migrant workers from the play or, generally, knowledge production on the valley: they had never thought about this absence as a problem, nor even realized there was in fact an absence. Thomas, who was so enthusiastic about the play's potential to gather the community after the divisive cable car, simply shrugged his shoulders when I asked him what he thought about what Ines called an 'erasure from history'. The latter's invisibility in the village's history was a non-topic for him, just like their non-involvement in Swiss and local forms of democracy was 'normal and good', to avoid the influence of foreigners who 'just' worked in the valley, even if they had done so for a long time. As Franziska acknowledged, too, the presence of these migrant workers, if economically central for the valley, was minimal in social and cultural terms: multiculturalism came from the tourists, not the population. The implicit local belief was that migrants would all leave again – for retirement, the low-season, or another job elsewhere – as soon as their activities were over.

What I initially interpreted as migrant workers' indifference towards tourism development in fact resulted from a locally and nationally crafted politics of difference. One could see such a thing as a direct, practical consequence of migration: for those moving from a different country, with its imagined community and temporal concerns (Anderson 2006), into an unfamiliar context often means being confronted with a different space and time. The examples I have provided here nonetheless show that the exclusion of migrants rests less on people's own, individual capacities to identify with a place, its history and future, than on an enduring system of (spatio-) temporal 'othering' of migrants. This othering consists of the inclusion of a migrant workforce in a national system, and their presence in a labour market, as well as the particularly excluding conditions in which this takes place. The integration chronotope that has shaped Switzerland's attitude towards immigration fuels a system that converts migrants into temporary, guest-like populations, and hinders their identification with national and local temporal frames.

What emerges from this is a highly contrasted picture of dependency. Where native locals temporally reasoned along their tourism dependency by virtue of their status as natural, primary subjects of tourism dependency, migrant workers were stripped of this possibility and remade into subjects of an instrumental, artificial, temporary and 'new' tourism dependency. But this did not just impact temporary, seasonal migrants and holders of the L permit, who sometimes spent several months a year in their home countries. The main narrative in the village and in Switzerland still held that migrants had to be grateful to even be allowed to work in the country, or else leave if the conditions did not suit them – just like guests, to copy Dušan's wording. Andrea, for instance, insisted that she had no other choice but to come to the valley because of the precarious economic situation in Portugal but that she was a foreigner and was subjected to difficult working conditions that she would not accept at home. Talking about the Swiss who refused to occupy positions similar to hers, she said, 'I would not want to do this work either if I were in my country'. After I asked why she accepted it here, she retorted: 'It's not my country, I came to work. I came to work. They haven't called me, but I came to work'. Voicing her negative experience of dependency, by for instance complaining, would, she believed, lead to her expeditious firing and replacement by another migrant worker, and eventually maybe even her deportation.

Framed as 'winners' or 'profiteers' of the system, or as liberal subjects who had chosen to come to the valley and had to be grateful for this opportunity, migrant workers were turned into artificial, voiceless tourism dependents. As De Genova and Peutz describe, the conditions of migrant workers or 'free' mobile labour are in fact based on 'a freedom that is distinctly circumscribed' (De Genova and Peutz 2010: 56). This derives from a reality where 'the "freedom" to move about and sell one's labour is produced by the lack of freedom to withhold one's labour' (Seiger et al. 2020: 4). Migrant workers I met all reported they had come because they had no other choice.

The interlocutors' accounts I have presented in this chapter illustrate how dependency participates in identity politics and the unevenly distributed capacity to make claims as tourism dependents (Ferguson 2013). The case of migrant workers – highly reliant on tourism but not legally, socially or affectively acknowledged as such – reveals that the (in-)capacity to express one's status as a dependent does not always correlate with actual experiences of dependence. In the next chapter, I will further unpack what such an experience entails for migrant tourism dependents, but to conclude, let me come back to Ines's words regarding the invisibility of migrants.

CONCLUSION

Shortly after Ines had told me about feeling invisible as a migrant worker in the eyes of locals, the Swiss, tourists – 'everyone', as she summed it up – I questioned her on the reasons behind such a statement. She clarified, 'It's not that they don't see us but they don't *want* to see us'. With this, Ines echoed Axel Honneth's description of the feeling 'of being made invisible'. This is a process 'of being made to disappear, that evidently involves not a physical non-presence, but rather non-existence in a social sense' (Honneth 2001: 111). For Honneth, this highlights the '*deformation* of the human capacity for perception with which recognition is connected' (ibid.: 126, italics in original). Ines did not know exactly what reasons motivated this attitude from the side of her fellow villagers but she was certain of its effects: making migrant workers invisible maintained the illusion that '[the Swiss] did everything by themselves and not [with the help of] the others'.

Not acknowledging the presence of migrant workers, in the play for instance, but also more generally in visual representations of the valley, had indeed the effect of presenting a form of 'spectacle' of a community – a smooth, romanticized and thus marketable image of Switzerland, cleared from the experiences of social stigma and exploitation that have shaped its history.[27] It overshadowed the fact that migrants like Ines had participated in maintaining such imaginaries via their labour, by for instance working for hoteliers who presented an idealized image of the place in their hotels, or by wearing uniforms that made them match the visual culture of the place. Moreover, the labour of the complex community in the present, made up of people from a variety of horizons, was used to create a fixed, pure, simplified image of the place, to make a spectacle of the community. This imaginary is not without consequences: as we have seen, it convinced even other locals, for whom migration was a reality or a 'problem' of urban areas and not of the valley, in spite of its importance.

If from a 'presentist' point of view, time cannot be accessed other than through the representations social actors make of it in the present, this chapter reveals how evocations of the village's long history and future – what could be termed non-migrant time – were indeed only possible thanks to those living in the migrant-time. For native locals to be able to live in their valley as original tourism dependents, and to express themselves as such, representing the valley's past and future required the importation of other dependents, migrant workers, who then became intermediary, unaffected casualties of the business, whose labour and presence could easily go 'unrecognized', as Ines said. Native natural dependency was predicated on the existence of a migrant dependency, framed as artificial, temporary and new.

In a place as dominated by Romantic imaginaries as the Alps, drawing on these relationships of interdependence is essential to providing a fuller, differentiated account of tourism dependency, along with its consequences and uneven manifestations. Yet, in a nativist *and* capitalist system, natives who relied on the business of foreigners, without whom they would be 'nothing', seemed to overlook the fact that they would not be able to maintain this industry without a foreign workforce either. This, again, reveals the limitations of seeing native locals solely as 'victims' of (over)tourism. Not only did they occupy a position of privilege over non-Western mass tourists, but they also relied on the outsourcing of labour to a migrant workforce.

NOTES

1. The Portuguese 'community' was often referred to by my Swiss interlocutors, who believed there were strong links tying the Portuguese together in the village. Contrary to what others have reported on elsewhere in the Swiss Alps (Graf 2021; Friedli and Boscoboinik 2023), there were no specific Portuguese events or clubs reuniting them. A little shop selling Portuguese specialties on the outskirts of the village was the only place where some – mostly men – would meet regularly. Unfortunately I only had a very limited access to this place.

2. All derived from the Swiss *Sonderfall* myth. The idea of Switzerland as an exceptionally wealthy country within Europe (but outside the EU) is another variation of the myth of exceptionalism (see also Eberle and Imhof 2007).
3. In the valley, some Swiss women were also active in the hospitality industry. Nonetheless, these women preferred to work in places (like Alpine huts) that were spared of the so-called mass tourists and were 'off the beaten track'. There, salaries were higher and working hours shorter, and the respect from European, individual and regular guests greater, as I heard from one of these workers. For many of these women, working in restaurants remained a secondary activity.
4. Ines only knew of one Portuguese who managed to buy something in the valley, but in a bad location – a shady area, next to the railway and the road.
5. *Überfremdung* also played an important role in Germany and Austria, but it became particularly prominent in Switzerland (Kury 2003).
6. My own translation from French.
7. The minimum residence requirement for a permanent residency permit at the time was ten years. The remainder consisted of annual permits (Fischer and Dahinden 2016: 10).
8. During the three-circle policy, the first and closest circle was made up of EU nationals, the second of US and Canadian nationals and the third of 'other' nationals. With the bilateral agreements between Switzerland and the European Union in 2002 on the free movement of individuals (AFMP), the three-circle policy became a 'two-circle policy', granting Europeans the same living and working rights as Swiss nationals, with the exception of the right to vote (Riaño and Wastl-Walter 2006: 1703).
9. See Jaberg and Duc-Quang 2017 and Duc-Quang, Scuderi and Willemin 2023.
10. The hospitality industry requires a significant amount of human labour relative to the capital invested. In the context of the hospitality sector, labour intensity translates to a high demand for workers, particularly those willing to perform physically demanding tasks for relatively low wages. The fluctuating nature of labour in this industry typically comprises fluctuations within the day, as well as throughout the year, with the high and low season. See Kalbermatter 2020; Iannuzzi 2023.
11. The other being the 'postmigrant' chronotope, which highlights the migration heritage of second-generation migrants in various political projects (Çağlar 2018). This chronotope is a more urban one, and is virtually absent from regions like the Bernese Highlands.
12. Often described as one of the toughest immigration regimes in Europe by experts (e.g. Bauböck et al. 2013; Schmid 2023) and the media (e.g. Swissinfo 2004).
13. The law differentiates between the 'regular naturalization' and 'simplified naturalization', the latter being accessible to foreign spouses of Swiss nationals (this, however, excludes registered partnerships) and third-generation immigrants. The conditions to access this simplified naturalization are five years of residency and 'sufficient levels of integration', such as language proficiency (Citizenship Ordinance, LN Art. 6). Given the federal system in Switzerland and the application of laws on a cantonal and municipal level, it is important to note that the treatment of the status of foreigners is marked by great discrepancies between regions. The federal law dictates a framework that the cantons and municipalities can implement differently. Some cantons and municipalities (the more rural) are more restrictive, while others (the urban) are more liberal regarding foreigners (Probst et al. 2019).

14. One of the main changes of this federal law has been a reduction of the time required to be eligible for citizenship (ten years instead of twelve). However, a particular emphasis has been put on markers of integration such as language with levels of proficiency required to pass the citizenship test (B1 oral and A2 written, according to the Citizenship Ordinance LN Art. 6), while the previous law contained no clear indication.
15. According to the Swiss State Secretariat for Migration, the conditions for family reunification are described as follows: 'Anyone wishing to bring his/her family to Switzerland must have suitable accommodation. An accommodation is deemed suitable if it reflects typical living conditions of Swiss citizens' (State Secretariat for Migration 2017).
16. According to Bakhtin: 'In the literary artistic chronotope, spatial and temporal indicators are fused into one carefully thought-out, concrete whole. Time, as it were, thickens, takes on flesh, becomes artistically visible; likewise, space becomes charged and responsive to the movements of time, plot and history' (Bakhtin 1981: 84).
17. As mentioned already in Chapter 1 and note 3, the traditional kin structure was bilineal, and it relied on the integration of outsider women through marriage, as in many regions of the Alps (Heady 1999). While men stayed and worked on their farms or their *Heimat* in the original sense of the term, their wives had often come from 'elsewhere' (usually other parts of Swiss-German Switzerland). Initially attracted by the tourism and the hospitality industry, many of these women had arrived in the village by the 1980s, worked on the family farms, married and raised their children, but they often worked for a couple of hours in the valley's surrounding huts and restaurants.
18. As locals reported, these were rare, but when they happened they caused memorable unrest in the village.
19. The differences between children claiming to be descendants of the original Swiss Confederates and other children with migrant backgrounds have given rise to some tensions in the Swiss context (Fossgreen 2015).
20. According to information they communicated to me in March 2020.
21. Such as the right to change one's employer, the right to family reunification and the right to unemployment benefits.
22. Only after completing a total of twelve months employment in the previous two years in Switzerland did these workers have the right to go to the Swiss unemployment office during the low seasons, where they would receive 70–80 per cent of their salaries. Importantly, the L permit does not constitute a pathway to citizenship as it does not count as time formally spent in Switzerland.
23. It is also exclusionary on other accounts. In terms of age first, the legal age of majority was defined according to the 'intellectual maturity' of voters. It also restricts the vote to the first residents (of the municipality) of three months to ensure 'real' residency/impactfulness as opposed to 'electoral tourism' of those who do not plan to live in the valley but wish to change its political course. These conditions have been subjected to debate in recent years but they have not had the same implications as the exclusion of voting rights for foreign nationals. Various commentators have deplored the fact that two million people cannot participate in shaping the future of their place of residence. See, for instance, the project '*Aktion Vier Viertel*': https://www.aktionvierviertel.ch/fr/arguments/ (last accessed 24 May 2024). On the 'electoral tourism' debate, see https://www.rrgr-service.apps.be.ch/api/gr/documents

/document/e3124f76adcd44baa11e3605b18f1c5d-332/6/M-101-2010-Vorstossantwort-fr.pdf (last accessed 24 May 2024).

24. For an overview, see https://www.ekm.admin.ch/ekm/fr/home/staatsbuergerschaft-citoyennete/Citoy/buergerrechte/panorama.html (last accessed 08 July 2024).
25. The Second Home regulation (ZWG), passed in 2016, and the construction ban, was said by many to increase property prices generally, especially for a municipality like the valley, whose percentage of holiday homes exceeded that dictated by the law (20 per cent). See Chapter 1, note 7.
26. In 2017, the mayor told me of cases of native outmigration he knew of as a problem faced by the village, but without drawing on concrete numbers. Municipal workers told me they were unable to respond to my queries regarding the demographic movements of the 'natives'.
27. I hereby rely on Gilles Debord's theory of the spectacle as a reification of commodities, which consumers observe passively. In this case the commodity is the touristic imaginary of the village community as an authentic Swiss village inhabited by natives (Debord 2010).

REFERENCES

Adler, Patricia A., and Peter Adler. 2004. *Paradise Laborers: Hotel Work in the Global Economy*. Ithaca, NY: Cornell University Press.

Afonso, Alexandre. 2006. 'Les métamorphoses de l'étranger utile. Internationalisation et politique d'immigration dans la Suisse du tournant néolibéral', *A Contrario* 4(1): 99–116.

Amarelle, Cesla. 2010. 'Les migrations économiques sans statut légal: L'environnement normatif des sans-papiers', in Cesla Amarelle and Minh Son Nguyen (eds), *Migrations et économie, L'accès des étrangers et la vie économique: Les normes et leur application*. Berne: Stämpfli.

Anderson, Benedict. 2006. *Imagined Communities: Reflections on the Origin and Spread of Nationalism*. London: Verso Books.

Arlettaz, Silvia. 2012. 'Saisonniers', *Dictionnaire Historique De La Suisse*.

Bakhtin, Mikhail M. 1981. *The Dialogic Imagination: Four Essays*. Edited by Michael Holquist. Austin: University of Texas Press.

Bauböck, Rainer. 1998. 'Sharing History and Future? Time Horizons of Democratic Membership in an Age of Migration', *Constellations* 4(3): 320–45.

Bauböck, Rainer, et al. 2013. 'Access to Citizenship and Its Impact on Immigrant Integration: European Summary and Standards'. Edited by the European University Institute and the Robert Schuman Centre for Advanced Studies.

Berliner, David. 2012. 'Multiple Nostalgias: The Fabric of Heritage in Luang Prabang (Lao PDR)', *Journal of the Royal Anthropological Institute* 18(4): 769–86.

Binford, Leigh. 2013. *Tomorrow We're All Going to the Harvest: Temporary Foreign Worker Programs and Neoliberal Political Economy*. Austin: University of Texas Press.

Brettell, Caroline. 2003. *Anthropology and Migration: Essays on Transnationalism, Ethnicity and Identity*. Walnut Creek, CA: Altamira.

Burawoy, Michael. 1976. 'The Functions and Reproduction of Migrant Labor: Comparative Material from Southern Africa and the United States', *American Journal of Sociology* 81(5): 1050–087.

Bryan, Catherine. 2018. '"Wait and While You Wait, Work": On the Reproduction of Precarious Labor in Liminal Spaces', in P. Barber and W. Lem (eds), *Migration, Temporality and Capitalism: Entangled Mobilities across Global Spaces*. Cham: Palgrave Macmillan, pp. 123–39.

Çağlar, Ayşe. 2013. 'Locating Migrant Hometown Ties in Time and Space: Locality as a Blind Spot of Migration Scholarship', *Historische Anthropologie* 21(1): 26–42.

———. 2018. 'Chronotopes of Migration Scholarship: Challenges of Contemporaneity and Historical Conjuncture', in P. Barber and W. Lem (eds), *Migration, Temporality and Capitalism*. Cham: Palgrave Macmillan, pp. 21–41.

Chambers, Erve. 1997. *Tourism and Culture: An Applied Perspective*. Albany: State University Press of New York.

Dahinden, Janine. 2016. 'Switzerland', in *The Wiley Blackwell Encyclopedia of Race, Ethnicity and Nationalism*, edited by J. Stone et al. Chichester: John Wiley, pp. 1–4.

Dauvergne, Catherine, and Sarah Marsden. 2014. 'The Ideology of Temporary Labour Migration in the Post-Global Era', *Citizenship Studies* 18(2): 224–42.

Debord, Guy. 2010. *Society of the Spectacle*. Reprint Edn. Detroit: Black & Red.

De Genova, Nicholas. 2019. 'Migration and the Mobility of Labor', in *The Oxford Handbook of Karl Marx*. Oxford: Oxford University Press, p. 425–40.

De Genova, Nicholas, and Nathalie Peutz. 2010. *The Deportation Regime: Sovereignty, Space and the Freedom of Movement*. Durham, NC: Duke University Press.

Duc-Quang, Nguyen, Fanny Scuderi and Antoine Willemin. 2023. 'Population, Immigration, Vieillissement: Découvrez l'évolution de la Suisse en 15 Graphiques', *Le Temps*. Retrieved 15 April 2025 from https://archive.letemps.ch/archive/www.letemps.ch/grand-format/population-immigration-vieillissement-decouvrez-levolution-suisse-15-graphiques.html.

Eberle, Thomas S., and Kurt Imhof. 2007. *Sonderfall Schweiz*. Zurich: Seismo.

Fabian, Johannes. 1983. *Time and the Other: How Anthropology Makes Its Object*. New York: Columbia University Press.

Falk, Francesca. 2022. *Der Schwarzenbacheffekt. Wenn Abstimmungen Menschen traumatisieren und politisieren*. Zurich: Limmat.

Ferguson, James. 2013. 'Declarations of Dependence: Labour, Personhood and Welfare in Southern Africa', *The Journal of the Royal Anthropological Institute* 19(2): 223–42.

Fibbi, Rosita. 2011. 'L'évolution des droits politiques des étrangères et des étrangers en Suisse: Quelle signification du point de vue de la citoyenneté?' *SFM Discussion Papers* 25: 16 Pp.

Fibbi, Rosita, et al. 2010. *Die portugiesische Bevölkerung in der Schweiz*. Bern: BBL, Vertrieb Bundespublikationen.

Fischer, Carolin, and Janine Dahinden. 2016. 'Changing Gender Representations in Politics of Belonging: A Critical Analysis of Developments in Switzerland', *Working Paper Series – NCCR on the Move*.

Fossgreen, Anke. 2015. 'Edelweisshemd sorgt an Schule für Eklat', *Tages Anzeiger*, 13 December. Retrieved 18 April 2025 from https://www.tagesanzeiger.ch/Edelweisshemd-Sorgt-an-Schule-Fuer-Eklat-223431154148.

Foster, Jason. 2012. 'Making Temporary Permanent: The Silent Transformation of the Temporary Foreign Worker Program', *Just Labour* 19. Doi: 10.25071/1705-1436.24.

Friedli, Andrea, and Andrea Boscoboinik. 2023. 'Mobilities in the Swiss Alps: Circulation and Rootedness', *Quaderns de l'Institut Català d'Antropologia* 38(2): 249–63. Doi: 10.56247/Qua.418.

Glick Schiller, Nina. 2009. 'A Global Perspective on Transnational Migration: Theorizing Migration without Methodological Nationalism', in Rainer Bauböck and Thomas Faist (eds), *Diaspora and Transnationalism: Concepts, Theories, Methods*. Amsterdam: Amsterdam University Press, pp. 109–29.

Graf, Flurina. 2021. *Migration in den Alpen*. Bielefeld: Transcript-Verlag.

Heady, Patrick. 1999. *The Hard People: Rivalry, Sympathy and Social Structure in an Alpine Valley*. Abingdon: Routledge.

Holmes, Seth. 2013. *Fresh Fruit, Broken Bodies: Migrant Farmworkers in the United States*, Vol. 27. Berkeley: University of California Press.

Honneth, Axel. 2001. 'Invisibility: On the Epistemology of "Recognition"', *Aristotelian Society Supplementary Volume* 75(1): 111–26.

Iannuzzi, Francesco E. 2023. 'Local Labour Market Segmentation and Migrant Workers' Experiences: The Case of the Hotel Industry in Venice', *European Urban and Regional Studies* 30(2): 172–85. Doi: 10.1177/09697764221145357.

Ioannides, Dimitri, and Kristina Zampoukos. 2018. 'Tourism's Labour Geographies: Bringing Tourism into Work and Work into Tourism', *Tourism Geographies* 20(1): 1–10. Doi: 10.1080/14616688.2017.1409261.

Jaberg, Samuel, and Nguyen Duc-Quang. 2017. 'Foreigners Arriving in Switzerland More Likely to Be Highly Skilled', *Swissinfo*. Retrieved 18 April 2025 from https://www.swissinfo.ch/eng/business/employment-and-migration_foreigners-arriving-in-switzerland-more-likely-to-be-highly-skilled/43606836.

Jain, Rohit. 2020. 'Schwarzenbach geht uns alle an! Gedanken zu einer vielstimmigen, antirassistischen Erinnerungspolitik', *INES*, 22 November 2020.

Janta, Hania, et al. 2011. 'Migrant Relationships and Tourism Employment', *Annals of Tourism Research* 38(4): 1322–43.

Janta, Hania, and Adele Ladkin. 2024. 'The Labour Consequences of Covid-19: Migrant Workers in Tourism and Hospitality', *Tourism Review* 79(1): 266–70.

Jordhus-Lier, David, and Anders Underthun. 2014. *A Hospitable World?: Organising Work and Workers in Hotels and Tourist Resorts*. New York: Routledge.

Kalbermatter, Jacqueline. 2020. *Bleiberecht in der Gastro-Küche: Migrationspolitische Regulierungen und Arbeitsverhältnisse von Geflüchteten mit unsicherem Aufenthaltsstatus*. Zurich: Seismo.

Kasmir, Sharryn, and August Carbonella. 2014. *Blood and Fire: Toward a Global Anthropology of Labor*, Vol. 13. New York: Berghahn Books.

Klimt, Andrea C. 1989. 'Returning "Home": Portuguese Migrant Notions of Temporariness, Permanence and Commitment', *New German Critique* 46: 47–70.

Kury, Patrick. 2003. *Über Fremde reden: Überfremdungsdiskurs und Ausgrenzung in der Schweiz 1900–1945*. Zurich: Chronos.

Ladkin, Adele. 2014. 'Labor Mobility and Labor Market Structures in Tourism', in Michael Hall (ed.), *The Wiley Blackwell Companion to Tourism*. Hoboken, NJ: Wiley, pp. 132–42.

Lamont, Michèle, and Virág Molnár. 2002. 'The Study of Boundaries in the Social Sciences', *Annual Review of Sociology* 28(1): 167–95.

Landwehr, Dominik. 2020. 'The Schwarzenbach Initiative'. Retrieved 18 April 2025 from https://blog.nationalmuseum.ch/2020/06/schwarzenbach-initiative/.

Leitenberg, Danaé. Forthcoming. 'Navigating Tourism: Mobility, Uncertainty and Inequality in the Swiss Alps'. *Mobilities.*

Lems, Annika. 2022. *Frontiers of Belonging: The Education of Unaccompanied Refugee Youth.* Bloomington: Indiana University Press.

Maiolino, Angelo. 2012. 'Die "Tschinggen" und die Schwarzenbach-Initiative. Von der Politik der Marginalisierten zur Mediterranisierung der Schweiz', *Terra Cognita* 2: 20–22.

———. 2020. 'Für Hunde Und Italiener Verboten', Swissinfo. Retrieved 18 April 2025 from https://institutneueschweiz.ch/Fr/Blog/209/Fr_Hunde_und_Italiener_verboten_Ein_Gesprch_mit_Angelo_Maiolino.

Marres, Noortje. 2012. *Material Participation: Technology, the Environment and Everyday Publics.* Basingstoke: Palgrave Macmillan.

Mcdowell, Linda, Adina Batnitzky and Sarah Dyer. 2007. 'Division, Segmentation and Interpellation: The Embodied Labors of Migrant Workers in a Greater London Hotel', *Economic Geography* 83(1): 1–25.

Messerli, Paul. 1987. 'The Development of Tourism in the Swiss Alps: Economic, Social and Environmental Effects Experience and Recommendations from the Swiss MAB Programme', *Mountain Research and Development*: 13–23.

Moser, Patrick. 1997. *'So wird die Jungfrau zur Demoiselle Gemacht': Projektierung und Bau der Jungfraubahn.* Zurich: Chronos.

Niederberger, Josef M. 2004. *Ausgrenzen, Assimilieren, Integrieren: Die Entwicklung einer Schweizerischen Integrationspolitik.* Zurich: Seismo.

Petite, Mathieu. 2013. 'Les montagnards face aux écolos. Une énième déclinaison au travers de l'initiative Weber', *Journal of Alpine Research* (Hors-Série).

Piguet, Etienne. 2013. *L'immigration en Suisse: Soixante ans d'entrouverture.* Collection Le Savoir Suisse.

Piguet, Etienne, and Hans Mahning. 2000. 'Quotas d'immigration: l'expérience Suisse', *Cahiers de Migrations Internationales* 37: 1–40.

Probst, Johanna, et al. 2019. 'Marges de manœuvre cantonales en mutation: Politique migratoire en Suisse'. Université De Neuchâtel.

Riaño, Yvonne, and Doris Wastl-Walter. 2006. 'Immigration Policies, State Discourses on Foreigners, and the Politics of Identity in Switzerland', *Environment and Planning: Economy and Space* 38(9): 1693–1713.

Sandoz, Laure. 2019. *Mobilities of the Highly Skilled towards Switzerland: The Role of Intermediaries in Defining 'Wanted Immigrants'.* Basel: Springer.

Schmid, Samuel. 2023. 'Democratising Switzerland: The Significance of the New Naturalisation Initiative', *EUI Global Citizenship Observatory,* Florence: Globalcit, https://globalcit.eu/democratizing-switzerland/.

Seiger, Fiona-Katharina, et al. (eds). 2020. *Migration at Work: Aspirations, Imaginaries & Structures of Mobility,* Vol. 5. Leuven: Leuven University Press.

Skenderovic, Damir. 2007. 'Immigration and the Radical Right in Switzerland: Ideology, Discourse and Opportunities', *Patterns of Prejudice* 41(2): 155–76. Doi: 10.1080/00313220701265528.

State Secretariat for Migration (SEM). 2017. 'Family Reunification Fact Sheet'. Switzerland.

Starobinski, Jean. 1966. 'The Idea of Nostalgia', *Diogenes* 14(54): 81–103.

Swissinfo. 2004. 'Foreigners Face Tough Road to Swiss Citizenship.' *Swissinfo*, 8 September. Retrieved 18 April 2025 from https://www.swissinfo.ch/eng/life-aging/foreigners-face-tough-road-to-swiss-citizenship/4086156#:~:text=At%20a%20national%20level%2C%20there,12%20and%2020%20count%20double.

van den Berghe, Pierre L. 1992. 'Tourism and the Ethnic Division of Labor', *Annals of Tourism Research* 19(2): 234–49.

Vuilleumier, Marc. 2012. 'Etrangers', *Dictionnaire Historique De La Suisse.*

Wicker, Hansruedi. 2003. 'Einleitung: Migration, Migrationspolitik und Migrationsforschung', in H.R. Wicker, R. Fibbi and W. Haug (eds), *Migration Und Die Schweiz*. Zurich: Seismo, pp. 12–62.

Zampoukos, Kristina. 2018. 'Hospitality Workers and the Relational Spaces of Labor (Im)Mobility', *Tourism Geographies* 20(1): 49–66.

CHAPTER

6

AN INDUSTRY WHERE ONE CANNOT GROW OLD

'How are you? Are you happy (*glücklich*)? Everything stable (*Alles stabil bei dir*)?', Rachid often asked me to begin our conversations on the road that led to my apartment. He had moved into the village a couple of weeks before me, following a decade of living in hotel rooms in Germany and Austria, after he had left his homeland Tunisia. Rachid was hopeful when I met him. In his studio, he explained that he felt that things were going to turn out fine for him, now that he lived in the 'beautiful village' and finally had a place of his own. Now, it was only a matter of time until he would be able to meet someone, marry and build a family in the valley. He dreamt of opening a small shop, having a business of his own, and being 'his own boss' because like his father, a shop owner in Tunisia, and his brother, who had made it in France, he had the soul of a businessman. These dreams seemed far from Rachid's present reality, but he believed they were all 'normal' wishes, if not needs, at the age of forty.

When we met for tea in his studio, Rachid repeatedly showed me around and stressed: 'See how I live, I am tidy, everything is in order'. His tidiness, he seemed to imply, pointed to his capacity to want, build and deserve more than 'just' a job as a waiter – what he had now. He could and would manage his life like his space. Working in the hospitality industry was 'just temporary' (*vorübergehend*), a necessary but exploitative fix that he had tried to leave multiple times, and came back to when he failed. Rachid had now been a waiter for two decades, and previously had only worked among domestic tourists in regional touristic resorts. The Alpine resort was the first touristic place of this size and importance, giving him the opportunity to work permanently. So now, everything would hopefully follow its 'normal' course.

As the weeks went by, Rachid's enthusiasm progressively faded. His job was hard, the team complicated, the food he was given at work unhealthy, and his days off were scattered throughout the week. The difficulties Rachid faced regarding his colleagues and work were not new but they had become 'unbearable' (*unerträglich*) – the kind of things that make you want to 'throw up' (*zum Kotzen*). 'I am fed up with it' (*Ich habe die Schnauze voll*), he regularly told me, throwing his hands into the air with irritation.

A small plastic bag that Rachid had left on his night desk symbolized the difficult situation he was in. 'Look at this bag; I've never taken so many pills in my whole life', he said with a mix of sadness and anger, pulling the coloured drug boxes out of the bag. Because of the stress, the work and the bad food, he developed a stomach ulcer. He had never been sick in his whole life, but now his body could not take it anymore. Rachid was worried he would have a burnout too, he told me, worried. Something was wrong with the hospitality industry, he said – so wrong that one could not grow old in it. He explained how few people over fifty were still working in the hotel that employed him, and how service work was essentially only made for young people. 'Reaching the retirement age is impossible', Rachid said. Receptionists and managers maybe, but not the unqualified, lower positions like his. He added, 'I have given up everything for this job but it has not given me anything'.

While Rachid had imagined a future with a wife, a family and a business of his own in the village, he was now unsure whether he should even make holes in the walls to install his TV: 'You never know, maybe I'll have to go'. But whilst he admitted feeling stuck, Rachid did not give up on his dreams at all: he still wished for a stable, happy life in the village, 'like everyone' (*wie jeder Mensch*). These dreams, even in the worst situations, continued to 'itch' (*kratzen*); they got under his skin and kept him moving, sometimes more, sometimes less. This is why, he explained, he absolutely needed to keep the studio, as his only and first safety net, a place of his own from which he felt he could move up towards his fantasized but universal dreams. Going somewhere else would mean starting from 'zero' again, leaving the place he had now made his own, and giving up on the dreams he had. To try and make these dreams come true, to pay rent, and to help his family at home with remittances, Rachid had to go back to work in a hotel – this one or another one made no difference.

The delicate relationship between hope and precarity, between tourism as simultaneously a means to establish a dream of stability, as well as a vehicle of uncertainty, is what this final chapter is about. Whereas I have looked at this issue throughout this book at a more collective level, in this chapter I will explore this tension on an intimate level, and specifically for migrant hospitality workers who have been left out of or erased from the long-term, national and local collective forms of temporal reasoning. If their temporal agency as denizens was limited, they remained important tourism dependents, with their own experiences of living off an exploitative but crucial industry to enter markets, national economies or dreams 'from below'. This chapter closely follows the ambiguity of tourism as a means to enhance or impede the better life *not* from the position of those who wish to stay put in a place that would be 'nothing without tourism', but from the position of those who count on tourism to move out of places marked by dispossession, precarity or crisis to become 'someone'. To explore these questions, I track the transnational dreams of stability and betterment around tourism as a pathway to a better life for Rachid, between his village in Tunisia and the Alpine resort in Switzerland. I then look at how this relationship to tourism and hospitality work as vehicles of existential mobility changed and contributed to Rachid's experience of feeling stuck as he grew older and his health deteriorated. While he dreamt of a normal life and believed that hard work would help him achieve it, he felt trapped in a difficult, corrosive and painful present.

TOURISM, MIGRATION AND THE TRANSNATIONAL PROMISE OF A BETTER LIFE

Rachid was born in Tunisia, in an area that greatly depended on tourism. Similar to the Swiss village and the Alps, tourism held strong promises of development for Tunisia during the colonial and postcolonial era (Saidi 2017; see also Hazbun 2008). In much the same way as tourism had enabled the Alps and the Bernese valley to thrive, it helped the Tunisian people make a living and escape poverty, Rachid said. In the 1960s and 1970s, tourism's creation of jobs and infrastructure contributed to the modernization of a new independent state (Saidi 2019). Tunisian seaside resorts built for middle-class French, German and British tourists made ideal tourist destinations, being in close proximity to Europe. Rachid, like many of his hometown friends, obtained his first real job as a waiter thanks to tourism – a booming industry that became synonymous with opportunities in a context shaped by dispossession and instability. Besides meagre revenues, the industry was also fertile ground for his dreams of a better life. In contact with 'Italian, German, French, and British tourists' that tourism had brought to the beaches near Rachid's home since a very young age – all 'really good people', he claimed – he began to aspire for more. He wanted to try his luck in Europe, like many of his fellow citizens and family members.

Because of Rachid's family's important role as translators during the French protectorate, but also because of tourism's postcolonial promises of development for his country, he often told me how, as a Tunisian, he naturally felt closer to Europe than to Africa. 'Tunisia is not Africa', Rachid regularly stated as he stressed his connections with the Arab and European worlds. Moroccans were 'brothers' and he was culturally more proximate to Europeans than the other Africans, he said, mobilizing a common, colonially inherited narrative that distanced the Maghreb from Africa (Perl 2020). But this belief was also rooted in his own experience, Rachid said, explaining how he felt even closer to Europe than most of his relatives and friends: 'My mother has always said I'm European in my soul'. Europeanness, Rachid implied, emanated from him, citing the example of a Belgian couple who 'wanted to adopt me when I was little, although I had a family'. The couple, he believed, must have 'felt' (*spüren*) his cultural proximity, even as a child. As he recounted the story decades after, Rachid was still half-flattered and half-confused by this event, but it unequivocally revealed the Europeanness that had shaped his life story from the very start, before he even migrated to the continent.

Leaving for Europe thus felt profoundly normal to Rachid, but it was also marked by difficulties. He recalled that he had left Tunisia because of a particularly unstable period. Despite his job and a completed training in the hospitality industry, he was unable to help his family and things deteriorated, under increasingly difficult economic and political conditions.[1] Rachid saw his parents and siblings suffer, and being unable to help his family made him feel ashamed and hurt. Gendered imaginaries of what a good life, a good son and a good man ought to be also informed these experiences: he should have been capable of supporting his family, someone that could eventually marry and build a family and a business of his own.[2] The similarities between Tunisia and Europe, the Europeanness that Rachid's family and tourists attributed to him, and the difficult present of dictatorship and financial struggles motivated Rachid to emigrate, like many of his compatriots (Brand 2006; Ragab, McGregor and Siegel 2013). As a national of a so-called third state according to the EU-Schengen policy, he applied for a student visa in Austria in 2000, where he wished to learn economics and get an education to make a better life than the one he had at home. Rachid described this as a normal trajectory – one he witnessed particularly in touristic regions where he worked and lived: 'Everyone wants to see something, to move, to travel – you too', he said to me, noting the number of tourists and retirees who came to Tunisia and the Alps.

The life Rachid imagined in Europe – one of greater efficiency, wealth, freedom and opportunities – nonetheless took a different turn when he went to Austria, having been accepted on a student visa. After two years of studying, he failed his exams and left university. Yet, Rachid could not imagine going back to Tunisia and its financial and sociopolitical difficulties under Ben Ali. During our discussions, Rachid often avoided talking about this part of his life, marked by failure and disappointment: he preferred to talk about how, thanks to hospitality labour, he was able to bounce back and stay in Europe under an immigration regime that would not allow him to remain without employment. He ended up going back into a sector that he had been involved since his young years and that 'was always hiring' waiters in restaurants and hotels in resorts in Austria and Germany.

Rachid recalled with a certain pride that, in spite of this change of plans, he enjoyed his life as a waiter and was always up for the next adventure. He was young, and the life that seasonal jobs in tourism offered was one of mobility, new experiences, travel and encounters. Rachid learnt new languages and skills, and enjoyed new people and places. He collected the postcards he received from his friends from all over the world, which decorated the door to his studio, as reminders of the good years, the meaningful times and experiences of the last decade or so. Since then, I too have contributed to this wall, sending Rachid postcards of the places I visited, as he asked me to, to enrich his postcard collection. Alternating between periods of hard work during the season and the long holiday – during which he visited his family and the friends he had met through work who lived all around Europe – was exhilarating, he told me. As the years passed, Rachid even made enough money to sit the Austrian citizenship test, which he passed.

TOURISM LABOUR AND MOVING FROM FUTURELESS TO FUTURAL PLACES

Although most of the migrant hospitality workers I met did not usually come from touristic places, their stories resonated with Rachid's. Andrea, Clara and Ines had felt compelled, even obliged, to leave their homelands because of incidents that had destabilized, interrupted or slowed down their lives. A present characterized by personal hardship, accidents, economic crisis, or dictatorship translated into their inability to imagine a future in their home countries. They migrated to (Central) Europe in quest of a future, for them and their families. Moving was a decision made under difficult circumstances, these migrant workers explained, but it was also a deeply normal endeavour: what nearly all their ancestors, their friends and their relatives had done before, and after, in order to live better. Clara, the Italian waitress briefly introduced in the previous chapter, often retold the story of her parents' migration from southern Italy to Germany to work. There, they faced racism, but worked hard to give their children a better future. Now, Clara described how she was 'doing the same' in Switzerland and wished to stay in the valley for her son to have a better future.

Narotzky and Besnier have shown how such temporal imaginaries are tied to migration as a spatio-temporal endeavour where hope consists of a 'material projection into a future that is located somewhere else' (2014: S11). For them, the affective force of 'geographical mobility' resides in the fact that 'it may translate into social mobility, it is hoped, in the right direction' (ibid.). At the same time, as Frances Pine points out, 'migration . . . can be both a symbol and an enactment of hope and of faith in the future, and an act of or a reaction to hopelessness, despair, and acute loss in the present' (Pine 2014: S96). This echoes the experiences of most of the hospitality workers I met, who identified migration as a necessity, as an exceptional response to a difficult situation, and as the manifestation of a universal, human and historical need to move forward and carve out a better life. Far from the clearly identifiable and external 'push and pull' factors so often evoked in the context of migration (De Haas 2011), Rachid's experience, like that of many of his colleagues, spoke of the interplay between physical movement and stasis on an existential level (Tošić and Lems 2019).

Ghassan Hage has most notably discussed the intertwining of the physical and existential movement with the concept of 'existential mobility' (Hage 2005, 2009, 2019). Based on his research among the Lebanese diaspora around the world and white supremacists, he has shown how the idea of 'a viable life' is linked to existential mobility, associated with the idea of 'going places in life' and 'going somewhere in life' (Hage 2009: 97; 2019). For him, the pursuit of existential mobility is usually or preferably done 'at home', in a familiar environment. It is only when his interlocutors felt that they were 'stuck' in their lives, that they were 'going nowhere' or 'too slowly', that they considered physical mobility: 'We move physically so we can feel that we are existentially on the move again or at least moving better' (Hage 2005: 470).

Similarly, moving to Austria, Germany and then Switzerland were life-enhancing pursuits for Rachid but also for his family. He highlighted the material and existential betterment brought about by the money he sent to his parents and siblings. This role felt reparative, not only because he helped to pay his family's medical bills and essential goods but also because, by taking his mother to Mecca – her dream – with money he had made in Europe as a hospitality worker, he had resolved the anomaly of being a powerless son, unable to look after his family. When his mother was depressed and sick, Rachid even managed to invite her to Switzerland, filling out dozens of complicated forms for her,[3] all aimed at 'lifting her up', he said, by coming to 'the beautiful village' when she was feeling down.

In both Tunisia and the Alps, tourism often appears to be a way to move forward in life, to achieve development for those who live in regions marked by dispossession or marginality. While this life enhancement often meant staying put in their valley for the *Einheimische*, Rachid's example shows how tourism labour also comprised an embodiment of the famous developmental narrative of tourism as a way to develop places but also people, contributing to their well-being and advancement. If tourism 'makes sense' for dependent populations precisely because of its short-term betterment of otherwise precarious conditions (Azcárate 2020: 8), analysts of mobile tourism labour have reveal that migrant workers view tourism and the labour opportunities it offers favourably when contrasted with a 'home' context shaped by hardships and instability (Adler and Adler 2004; see also Elgammal and Wilbert 2013; Saidi 2019).[4]

At the same time, Rachid's case points to how tourism's imaginaries (and realities) of cosmopolitan exchange and discovery greatly contributed to the emergence of his desire to move existentially via emigration. Tourism as an agent of development in postcolonial Tunisia, bringing mostly Western guests to the beaches near Rachid's home, made dreams of betterment possible for Rachid, revealing how tourism jobs 'offer contact with foreigners that can prove more valuable than their wages' (Brennan 2004: 62). Yet, it also did so because of the gap Rachid perceived (and was made to perceive) between his country and 'civilized, rich and free Europe'. Tourism and postcolonial power relations made his dreams 'dream-able' but also revealed the fundamental differences and inequalities between him and the guests. Rachid was unable to go to Europe as easily as Europeans (from France, Germany, Britain) came to Tunisia, as tourists, 'expats' or 'retirees'.

This unevenness shaped Rachid's time in Europe, too, and his return to hospitality work. Thanks to tourism labour, he was able to stay in Europe after having interrupted his studies. While this change of plans corresponded to Rachid's comeback to what he was professionally trained as, it was also a setback. It meant (re-)entering an economy that employed the 'unskilled' under difficult conditions and was always hiring, especially other migrants who 'did not have any other choice' but to work in precarious positions, as he often repeated, echoing with analyses of tourism labour (Janta et al. 2011). Andrea, the Portuguese migrant I introduced earlier, for instance, saw her experience and life path in Portugal made insignificant in Switzerland. Her university degree and experience as a trained nanny were made worthless as she migrated. Like her, many foreigners who came to the valley did have work experience, sometimes even in the hospitality sector, but were effectively stripped of these qualifications as they more or less had to start a life from scratch. Because they did not speak German well, because they had no previous experience in Switzerland, and because their diplomas were not considered worthy, they were in fact 'deskilled' (Barber and Lem 2018; Seiger et al. 2020) and paid accordingly in Switzerland. Crucially, such limited labour opportunities were seen as entry-level jobs in a wealthy country that, for many migrants, was conflated with possibilities regarding moving forward in life. In Rachid's case, Switzerland – and the Alpine village in particular – evoked the possibility for him to achieve a better salary and working conditions than in Austria and Germany, where he had only managed to get seasonal contracts.

In spite of their common aspirations to find stability and meaning in Switzerland, Rachid explained that migrants were treated far from equally. In the hospitality industry particularly, some nationalities had it easier than others in the country and in life Rachid said, echoing the sentiments of many others in the resort. In spite of his Austrian citizenship, Rachid felt different from the educated, young Germans who were, besides the Portuguese, another nationality that dominated the Swiss hospitality market. The Germans, he explained, had access to higher positions in the resort's and in Switzerland's economy because they benefited from better training, education, and language proficiency. Rachid, like other Portuguese or Italian workers I met, bitterly noted that the Germans had treated him superciliously, reminding him of his 'lower rank'. Rachid particularly resented the Germans, who 'controlled' and 'surveilled' him in his current position, but also because he had been discriminated against by them previously because of his Tunisian, non-European origins. Rachid had fewer problems with the Swiss, as he had very little contact with them, except as guests, he admitted. German hospitality workers I met in turn told me how they themselves felt disrespected by the Swiss employers and population. Sandrine, a receptionist from Stuttgart, explained how shocked she was by the poor treatment she had received from her Swiss employers in a four-star hotel. She noted, however, that she still considered herself lucky compared to Italian and Portuguese colleagues, who were treated with even less respect.

Such important differences in nationality among workers were an integral part of the local infrastructure of the hospitality industry, mirroring what has been described by many as a global feature of tourism labour (Adler and Adler 2004; McDowell, Batnitzky and Dyer 2007; Zampoukos and Ioannides 2011). Those working both inside and outside the hospitality industry told me how its 'frontlines' were typically occupied by those who spoke German fluently, which was important to convey an image of professionalism and proximity to local culture. The most culturally distant workers, such as Portuguese or Slovakians, were relegated to the 'backstage' of the hotels, performing non-interactive tasks such as washing dishes or working as chambermaids, and occupying the lowest positions on the market. While facts like the insufficient mastering of German or a lack of experience legitimized such distribution of labour, this was also aligned with particular culturalizing claims (see Chapter 5 and Kalbermatter 2020 for Switzerland). In such a competitive context, existential mobility was a highly individualized endeavour, about which Rachid, Ines and many others often complained. A sense of solidarity sometimes emerged from living together in workers' accommodations and from working together nearly every day during the high season, in the Swiss village like in other contexts (Jordhus-Lier and Underthun 2017). However, hospitality workers expressed a need to look after themselves in a merciless economy. The fluctuating, seasonal nature of the tourism economy and the temporary contracts led to a feeling of suspicion and jealousy among workers, who felt replaceable.

The critiques targeting workers from other nationality groups – often Germans – in the hospitality industry, uttered by Rachid and his colleagues did not simply reflect the widely reported, problematic and highly hierarchized features of the market (Duncan, Scott and Baum 2013). They also entailed a deeper critique of others' capacity to 'move better in life' – an aspect that Ghassan Hage has termed 'mobility envy' (Hage 2009: 98–101). Hage takes the example of mobility envy among white Australians towards refugees, but Rachid's accounts reveal that such feelings of resentment may exist within migrant populations too, especially when capitalism and nativism overlap to create a strongly segmented labour market. In his portrayal, Rachid implied that Germans not only started from a different position in life, but that they also moved better existentially and socially in an economy that acknowledged their skills and offered them prospects. Such mobility envy was not only directed at workers of a higher status. Ilana, a Portuguese hospitality worker and Ines's colleague, for instance, felt most resentment when she saw refugees – Eritrean, Afghan and then Ukrainian – arrive in the village and be given 'everything', contrary to her people. Her Ukrainian neighbours were the symbol of such unfair treatment, for whom they had been made the culprits. They were profiteers, whereas migrants like her had to struggle to earn the little they did.

AN OBSTACLE TO THE BETTER LIFE

When I first met Rachid, he did not hint at the sacrifices that hospitality work entailed for him. As he regularly stated with a smile, he approached the modest ideals of the normal life that he had set for himself in the village. Rachid believed that everything else would naturally follow after he had secured a permanent position and the studio of his own in a beautiful place. It was only three weeks into my fieldwork that Rachid suddenly revealed a radically different reality to the one he had been describing until then.

On that day, I met Rachid on his way home, when he was coming back from the hotel where he worked. A closed expression on his face, he spoke of his tiredness and sadness but mostly of the anger that he felt. He was outraged by his employer, and by his German superiors and colleagues, who were all too scared to speak up against the conditions they were subjected to. Rachid saw the sacrifices he had made for this industry, which had been unreciprocated, and this unfairness had become unbearable. Starkly contrasting with tourism development and work as a positively moving force, motivating him to leave Tunisia and to go to Europe as the place of a better future, Rachid now saw the tourism industry for what it really was: an exploitative industry, stealing from its foreign employees, and making it 'impossible to grow old'. 'This industry has no future', he repeated. Over the next months, until my (first) departure from the village in summer 2018, our discussions focused on his difficulties, which only seemed to be growing with time. As we met for coffee or tea, lunch or dinner, I witnessed Rachid's state deteriorate. He felt lonely and so was happy to have found an attentive listener in me. He also saw this as an opportunity to share his story, as it 'needed to be heard' in order for others to realize the truth about the industry.

But Rachid's own youth as a hospitality worker was also different from the bitter present. At the time, he was almost naively following every job opportunity, friendship and relation offered by the tourism industry. Like the European tourists he had hosted in Tunisia, Rachid explained that he had led a cosmopolitan lifestyle which was exhilarating, made up of excess and pleasure: women, alcohol, money. Work and leisure were interwoven in ways that were not dissatisfying as he worked intensively, under harsh conditions, but also 'had a lot of fun' and long, well-deserved vacations. Some of the youngest seasonal workers I met in the valley, active in the hospitality industry, in shops or ski schools, intentionally sought the same blurred blend of work and leisure that Rachid talked about. From Germany, Italy or the Czech Republic, these young workers sought the mountain climate and chance to ski and enjoy the winter or summer in the village *while* working, as reported by various tourism scholars (Bianchi 2000; Uriely 2001). This crowd was young and uncommitted, and enjoyed the momentary freedom.

It was only slowly, and two or three years before we met, that Rachid found he was becoming increasingly tired. The coming and going between places had worn down his capacity to feel like he belonged anywhere, as he told me. The hotel rooms where he lived felt more and more anonymous, and the endless goodbyes with friends weighed heavily upon him. It was as if the feeling of having 'no future', which had originally prompted his migration from Tunisia to Europe, had slowly come back to him when we met in the Alpine resort. Life again seemed to be 'not leading anywhere', in Hage's words (Hage 2005, 2009), and he deeply yearned for a 'stability' (*Stabilität*) that he felt his life lacked. Rachid now wished to stay put while remaining existentially mobile, moving upwards socially without having to move again, with all its associated problems. For Rachid, like many migrant workers but also native locals – as I have shown previously – it was thus precisely *not* moving that was seen as a sort of 'launching pad' (Hage 2005) for the better life (see also Leitenberg, forthcoming).

FEELING STUCK

With time, Rachid realized that his lifestyle as a seasonal worker and the physical mobility he enjoyed had become entrapping. Travelling everywhere around Austria and Germany, where the tourism industry would employ him, he was no longer going anywhere in life, nor near where he, his family and society had expected him to be by the age of forty. Rachid wanted to build a family, have an apartment of his own, be his own boss, all 'natural things', he kept insisting, but his job, its seasonality, the movement, the arguments with his colleagues, the running around, and the small hotel rooms were obstacles to his fulfilment. He thus moved again, to the Swiss Alps this time, hoping that it would be the last move as he had the promise of a first permanent position. If there was a place where his life path could be appreciated, he thought that it would be here, in this 'multicultural village', as he said – a place where he could make use of his German, French, English and Arabic when working with the tourists. But bad working conditions, the humiliation, the little free time and his body – his working tool – now turned against him, his gut quite literally telling him to stop.

In their work on hospitality workers in hotels at touristic resorts, Adler and Adler report on the feeling of 'entrapment' encountered by migrant hospitality workers from the Philippines[5] as the result of an aspiration: that of wishing to permanently settle on an island providing a limited 'local job pool' and thereby being exposed to the 'vicissitudes' of the market (Adler and Adler 2004: 6, 41). They note: 'Although their disposable income was limited, once they had paid their mortgages, car payments, and other expenses, and after they had sent money to their relatives back home, they could have saved for and made another move if they so chose. But they were happy in this poly-ethnic melting pot that embraced them' (ibid.: 41). Following Adler and Adler, many authors have reported on the feelings of 'entrapment' experienced by their interlocutors in the hospitality industry, especially those of migrant background (Alberti 2014; Jordhus-Lier and Underthun 2014; Bryan 2018; Guidetti, Pedrini and Zamparini 2021). Contrary to Adler and Adler, many have described how such experiences often came together with strong feelings of dissatisfaction for workers towards a situation of intersecting precarity, social immobility and discrimination shaping their employment (McDowell, Batnitzky and Dyer 2009; Alberti 2014). If Adler and Adler have shown how 'an objectively "bad" job can be perceived as subjectively "good" by workers', as Knox et al. put it (Knox et al. 2015: 1547), I believe that their interpretation of the willingness shown by migrant interlocutors to voluntarily 'entrap' themselves in Hawaii contains some limitation. For many, especially precarious workers, the financial obligations of paying off mortgages or sending remittances are not easily attainable, one-time objectives. Instead, these commitments often extend across a significant portion of people's lives. The experiences of Rachid, Ines and Clara in the Swiss Alps reveal a feeling of entrapment marked by a negative feeling of 'having no other choice' but to work in hospitality, precisely because of such financial concerns, as they themselves often expressed. This does not mean that they were always or just dissatisfied with their jobs, as Rachid's life story shows, but that these sentiments dominated.

Such entrapment manifested not only materially. Rachid was physically unable to move around as he wished with his ulcer, but his pain also pointed to a deeper, existential sense of being stuck. As Hage describes, stuckedness is 'a situation where a person suffers from both the absence of choices or alternatives to the situation one is in, and an inability to grab such alternatives even if they present themselves' (Hage 2009: 4). Everything hurt, inside and out, but Rachid stressed that he needed to work for his family, to whom he had to send remittances and maintain what he had achieved until now. Not working was no alternative: it would only make his dream of leading a better, normal life with a stable lifestyle even more out of reach. He could hypothetically leave the village for a better job in the urban lowlands but doubted that he could access any of those outside of the hospitality industry. Besides, he could not imagine letting go of this studio, the first of his own, and, it seemed to me, the life he had imagined in the village. In a touristic place shaped by the unequal division of labour between foreigners and Swiss people, finding a job in another hotel would be easy but the conditions were likely to be just as bad, he explained. In addition, the migration regime that I described in the previous chapter meant that, as a foreign national, Rachid needed to remain employed if he wished to stay in the valley, and indeed in Switzerland.

But the more Rachid worked, the worse his situation became. Following his ulcer diagnosis and during the treatment, he found himself in a difficult state where he was 'just working or recovering from work'. Rachid led what he described as a deeply destructive everyday, not only for his body but also his social life and his ability to 'meet someone'. Still, he did not know what else to do. The village doctor – a friend of his boss, Rachid painfully found out after his visits – refused to give him more than three days off in a row. He felt trapped in the industry and in the local structures of power and acquaintances. My conversations with him often concluded on a similar tone to those with Ines: 'We foreigners work in this industry because we do not have any other choice, contrary to the Swiss'. Rachid thought it was his duty to warn younger workers from Switzerland or Germany to leave the industry when he saw them doing internships at the hotel because it was not yet too late – for them.

One day, Rachid came back home with a smile. He had found a new job, one that was particularly rare as it would enable him to keep his evenings free. It was 'still' a job in the hospitality industry, a sector that he longed to leave, but maybe this time it was a good one. His body had now more or less recovered, and the manager – 'not a German', he specified – seemed to be a good man and appeared to take him and his experience seriously. Once again, Rachid felt closer to his goal: he was now able to spend evenings in the village, to chat with people and maybe even meet someone. The horizon of a stable life seemed reachable again. But, as in a bad dream, the promises of this new job quickly began to fade away and he found himself in a similar, if not worse, situation to his previous positions.

Within weeks, Rachid believed that a German colleague had become intimate with his boss, and he suddenly found himself stripped of the little authority and respect he had been given. The boss's new lover, a 'typical German' who thought she did everything better than him, decided she did not like Rachid for reasons he could not really understand. 'Because I took too much space or did too much', he thought. When he went to work, he promised himself to do only what was necessary and make himself as small as possible, so as not to upset her and get in trouble. After a few weeks, and in spite of these efforts, the boss told him he did not want him to come back. 'He said I didn't fit in with the team', said Rachid just a few hours after the event, outraged. As we went into his apartment, he explained that he had waited for me to pass by for hours, with his anger building up, in order to share the bitter frustration with someone and let off some steam. Nobody had ever treated him as badly in his long career. No one had dared to tell Rachid, with his joyful, loyal character, that he did not fit in with the team. All other employers could swear on their lives that he was good – one of the best employees they had ever had. 'I promise, they have only ever complimented me until now', Rachid repeated. He felt betrayed and disrespected again. 'I'm a human, I deserve basic respect', he said. The worst thing, Rachid noted, was that he had 'no one to talk to about this': his former co-workers, he feared, would laugh at him or gossip about his failure, and he did not have support there anyway. 'I told you this business has no future', Rachid concluded. I could not help him but listened to how he felt, nodding along.

Again, Rachid felt stuck, but he now firmly decided to not go back to the hospitality industry, no matter the cost. He went to the regional unemployment office and took the risk of losing his residency permit, his studio and everything he held dear.[6] This moment was the hardest of his life, he told me retrospectively. 'It took a lot of courage.' Rachid described how he had to literally 'resist' the temptation to work again under the same conditions, as he was obliged by the unemployment office to fill in a dozen job applications monthly for the same jobs in the hospitality industry. To add to his financial difficulties, his right to receive unemployment benefit was at first contested after he had been fired during his trial period. When he was later declared eligible, his unemployment money amounted to only 70 per cent of his salary. Rachid recalled how he 'changed his mind repeatedly within a day', how he switched from depressed to hopeful, from angry or sad to enthusiastic. He felt shame before his former colleagues, bosses and landlord, and felt as if he were walking on a tightrope, on the verge of depression and feeling alone. From having no time at all, Rachid explained how he suddenly had too much time, to think and worry, to question everything.

After months of unemployment, soon after I left the village Rachid was finally able to find a job as a salesman in a shop, thanks to a contact. Although he again described this new position as a luxury in a touristic village because he had normal working hours and an undetermined contract, he was still moved to do 'something more' than being someone's employee. This job, while much better, did not equate with 'the dream', but at least he had time to rest. When I saw Rachid in January 2020, he still yearned for a stable family life in the village, and said that his dream of being his own boss was 'still nagging at him'. A few months later, after the start of the Covid-19 pandemic, he said that this was not the moment to realize his dream and that he was now grateful to just have a job in a time of crisis, even if he had been temporarily laid off and was earning less than normal. At the same time, Rachid explained that this had further consequences because of the remittances he still had to send to his family. With no family or friends in the village, and without the chance to visit his relatives in Tunisia, the weeks of lockdown in spring 2020 were a particularly lonely time, even though he was in the 'beautiful village'. Maybe it was time to leave it altogether, but he could not risk it right now, he concluded.

THE PRESENT AS IMPASSE

Rachid's relationship with hospitality work and mobile labour as a means to achieve a better, more stable life thus slowly but dramatically changed over time. His age and the state of his sick body pointed to a form of deeper, existential crisis: he told me he had never been sick, never been ill, and had started 'not to feel himself anymore' because of the nature of the work he performed and the exploitative conditions to which he was subjected. These had all transformed Rachid from a joyful into a depressed and bitter person, as he sadly noticed himself, worried that he would one day have a burn-out, or worse. Rachid did not understand how state institutions could let things like this happen in a wealthy country like Switzerland, or why neither unions nor 'anything on the side of the workers' were not more present in the sector.[7] But he also had very little hope of change because the industry's very economic success depended on the fact that it could profit from, mistreat and replace workers easily, through the absence of state intervention and worker solidarity.

Rachid's profound awareness of the moment he was in resulted from its abnormal character: what was bearable and even enjoyable before, thanks to migration, tourism and hospitality work, had turned into a present that had to be endured, resisted and, as we will see, managed. From a sense of freedom embodied by many younger and often privileged temporary workers (Bianchi 2000; Adler and Adler 2004), he had slowly become entrapped in a position that he originally saw as temporary. Although Rachid's case was particularly dramatic, the sense of an abnormal present was something many migrant hospitality workers I met had in common. Nearly all suffered chronically from physical and psychological diseases – diabetes, lupus, hypertension, hernias, articulation problems, depression, burnout, etc. – that weighed on the everyday and were difficult to simply set aside. As abnormal as it felt, this was no exception for the village, or Switzerland even; it was part of a common reality for hospitality workers around the globe (Sánchez-Rodríguez et al. 2022). If these migrants were excluded from the valley's communal temporal reasoning, they also explained how difficult it was to extract themselves from the present as a result of their positions, where working conditions and precarity everyday shaped a grinding, and physically and emotionally painful present. Many of these workers lived in states of individualized, embodied 'enforced presentism', to coin Jane Guyer's famous concept: the present had an entrapping quality, that was, as in Rachid's case, often reinforced by expectations of a better future that seemed ever more distant, utopian and dreamlike as their states deteriorated (Guyer 2007: 410, see also Ringel 2018).

Rachid's experience of the enforced present had both a tangible and a mysterious quality. He was in search of explanations and justifications, but very aware of the situation in which he was living. On some days, I was surprised by the clarity with which he assertively reeled off facts and figures, having observed the mechanisms of the industry. On other days, he seemed lost, unable to understand why the hospitality business would dehumanize him in this way.

TOURISM DEPENDENCY AS CRUEL OPTIMISM

Rachid's experiences of tourism dependency, shared in many aspects by other migrant hospitality workers, radically contrasted with that of the Swiss and native locals I met during my research. Whereas the latter collectively reasoned towards the contested future with or without the cable car, while yearning for the 'better past' of farming or the 'good' guests and tourism, Rachid felt stuck in the present, unable to imagine a future for himself. Yearning for the past did not offer the 'palliative' and meaningful effects it had for other locals (Hodges 2010): going back in time (to life in Tunisia, or to his youth) meant embracing the conditions of precarity or instability that he had and still wished to escape. The long scope of temporal reasoning that characterized the Swiss and native villagers' dependency on tourism stood in radical opposition to Rachid's individual, embodied sense of 'stuckedness' in the present.

One eminent strand of scholarship has revealed how migration often translates into experiences of radical uncertainty and feelings of entrapment in the present, especially in a political conjuncture that imposes restrictions on the mobility of those who attempt to move from the Global South to the Global North (Griffiths 2014; Ramsay 2017; Khosravi 2021). Although experiences of radical uncertainty and waiting marked experiences of other migrants coming to the Alpine valley – as in the case of the Afghan refugees I was in contact with in the village, who were waiting to be granted refugee status or protection – Rachid's experience cannot be considered solely through the prism of migration as a vector of uncertainty. In what follows, I propose a reflection on the central role played by tourism dependency *and* migration in Rachid's feelings of stuckedness, as conditions favouring the creation and maintenance of a relationship, which Laurent Berlant has termed 'cruel optimism'.

Cruel optimism describes 'a relation [that] exists when something you desire is actually an obstacle to your flourishing' (Berlant 2011: 1). Berlant identifies the 'problematic object' of this relation as 'the affective attachment to what we call "the good life"' (ibid.: 27), meaning 'the fantasy of meritocracy, a fantasy of being deserving, and its relation to practices of intimacy, at home, at work, and in consumer worlds' (ibid.: 167). This points to a paradoxical situation where 'bad life . . . wears out the subjects who, nonetheless and at the same time, find their conditions of possibility within' these attachments to the good life (ibid.: 27). Such optimism is cruel precisely because it survives and invigorates individuals to continue to invest in, and often make sacrifices, to achieve the 'good life' in spite of the absence of basic conditions that could guarantee it. Worse, in working towards these utopias, people behave in ways that threaten their present well-being and their capacity to even achieve such ideals. As such, cruel optimism simultaneously disorganizes the life of those who experience it as well as organizes it: it produces meaning, movement and structure in the everyday while also wearing one out. The consequence of such affective attachment is that one 'might not well endure the loss of their object/scene of desire' (Berlant 2011: 4), because this relation provides individuals with a sense of stability, identity and their place in the world.

Relationships of cruel optimism flourish in particular contexts: besides those that Berlant identifies – austerity and shrinkage of the welfare state under neoliberalism, marked by 'crisis ordinariness' – I would add that capitalist tourism offers a particularly fertile ground for cruel optimism, even under its most historical, stable forms. In the Swiss resort and other tourist destinations, those who depend on this industry to make a living have seen in it the possibility to achieve and secure the good life, particularly so in marginal regions. But, as Azcárate argues, the betterment enabled by tourism rests on sacrificial logics in the present, because of its devastating effects on environments, bodies and relationships (Azcárate 2020; see also Brennan 2004). This was also what I could witness in the Alpine valley, with the evocations of a stable, secure and happy future, all leading to important sacrifices in the present on an individual or collective basis, such as accepting the tourist crowds or the important constructions in one's valley. The promise of a collective future also rested on a divided, sacrificed present. These debates, whether in the Bernese Alps or in the Yucatán, speak of the contemporary challenges of tourism as a sacrificial industry, which appears as a particularly powerful vehicle of cruel optimism for those who depend on it to exist.

This rings true for many of those who expressed their wishes to remain in the valley thanks to tourism, and particularly so for migrant hospitality workers like Rachid. Mobility and migration represent one of the axes of difference along which particularly cruel forms of optimism sprout. As Pettit and Ruijtenberg have argued, cruel optimism in migration contexts 'is not produced simply through "resilience" or "endurance", but by migratory regimes' (Pettit and Ruijtenberg 2019: 733). This system 'encourage[s] people to develop cruel attachments to objects that are embedded with the promise of migratory success, but at the same time inhibit the means of achieving it for many' (ibid.). Examples of lives of migrants yearning for stability in countries deemed 'better' than theirs, but confronted with the crushing reality of structural racism, administration or humiliation, indeed abound (De Genova and Peutz 2010; Lems 2019). In Switzerland, a country that has so often been portrayed as an *Eldorado* for migrants, but where foreigners have been confronted with the state's temporizing practices, migration represents a crucial site for relationships of cruel optimism to emerge. Being caught in a 'perpetual state of becoming' (Çağlar 2018: 27) points to a future and deferred state of stability and a present made of precarious social, affective, legal and economic conditions. For Rachid and many others, achieving the future of existential security that was 'promised' through migration, required enduring unequal socioeconomic and legal conditions imposed by the Swiss migration regime.

Rachid's life story captures how tourism labour and migration, in alternation or in concordance, both participated in his own, existential pursuit of the better, the 'stable' or the 'normal' life, as he called it. Tourism had played a central role in exposing him to dreams and conventional models of success, such as making it in Europe, the place of civilization and order, as he said. This then motivated Rachid to emigrate, to cross borders, as he had felt existentially stuck in Tunisia. When his ambitions did not go to plan, he went back to tourism as a migrant who 'had no other choice' but to work in this industry, to achieve his dreams of being his own boss, finding a wife and making a family, and owning a house. Tourism and migration had long structured and organized Rachid's everyday, and his sense of self. 'I am a boss, a businessman in my soul', he kept telling me, even though his position at the time was, he admitted, 'just that of a servant'. In this context, migration and hospitality work in Switzerland constituted an optimistic means to invest in a relationship of proximity to the good life, to attempt to realize these desires of stability from the bottom. But because of the particularly precarious working conditions pertaining to mobile tourism labour, the low prospect of upward mobility, the pressure to send money to his family, and the competition between him and his colleagues, Rachid led a life where the harsh present stood in tension with his dreams.

Stuck in the present, Rachid believed that he would lose either way: continuing to be employed in hotels and restaurants meant continuing to be subjected to destructive work, destabilizing conditions and the disturbing feeling that he could no longer recognize himself; leaving meant jeopardizing his future and the chance to become what he really wanted, deep down, as an ambitious man – namely, to be a protective husband and son, and to protect the future of his other relatives. He could not 'go on like this forever', not without losing his mental and physical health; but doing things differently seemed impossible, too. Rachid, however, was not deprived of agency as he eventually resisted the appeal of tourism labour as a vehicle to the better life, and left the hospitality industry to surpass the impasse in which he found himself, which he managed to do but at great emotional and financial cost.

Many of Rachid's fellow migrant workers similarly described *not* working in tourism as being as destructive as working in it, precisely because it signified their incapacity to achieve the better life. If they were to stop working as hospitality workers, their dreams and aspirations – which so deeply legitimized their migration and gave meaning to the harsh everyday – would disappear. As they so often repeated, they had migrated *for* their families, *for* their stability, and *for* a better future that would correspond to what they really were: a caring and sacrificing mother, a businessman or (more rarely) businesswoman, a future husband. Losing tourism labour meant losing a part of themselves. Their yearning for such objects of the good life cruelly led them to feel that their dreams were ever more distant as time went by. Both Andrea and Clara wished to be good, present mothers, but they could only spend a little time with their children in the present as they were working to give them a future. Rachid, at heart a joyful businessman, a dynamic man and a future fiancé, was lonely, tired and depressed as he worked to attain a stable life. Ines, who left Portugal to have a better, free and independent future, ended up in an entrapping, financially difficult present for many years, in an abusive relationship with a husband and co-worker. Only her brother, who had stayed in Portugal, was spared such a migrant reality and could remain 'true to himself'. Instead of working towards her dreams and living *off* tourism to achieve them, she told me that she was in fact 'living *for* tourism'.

Between the different types of labour they had to perform and their life more generally, many migrants felt like they had lost connection with their real self. Rachid explained that, besides the mental and physical labour he had to perform, such as counting, switching between languages and running around, he also had to sound, look and feel in accordance with the guests' and employers' expectations. These pressures stemmed from the widely reported consequences of 'tourism imaginaries' or 'tourist gazes' on local populations (even when they are not actually considered as such by the rest of the community in this case), such as requiring them to wear clothes that evoked typical Alpine imaginaries (Urry and Larsen 2011; Leite 2014). At the same time, these workers also had to make guests feel good, by being overly polite, friendly and smiling. The all-encompassing nature of the work put to respond to these expectations – a physical, intellectual, aesthetic (Witz, Warhurst and Nickson 2003) and emotional type of labour (Hochschild 2003) within the hospitality sector – added to the already harsh material conditions of the job, and they followed Rachid long after work. As he said, and as many others echoed, the stress he endured at work took time to come down.

Rachid and his colleagues described the necessity to develop what Bunten has described as a 'commodified persona', resulting from 'a strategy that cultural-tourism workers use to manage and control their product: themselves' (Bunten 2008: 390). Often, and particularly when work was stressful, this persona did not leave their bodies after the end of the day, but left a deeper mark: that of having become someone else (Hochschild 2003:

7; Brennan 2004; Bunten 2008).[8] If cruel optimism is part of the current, neoliberal constellation, Rachid's experience, between migration and tourism labour, points to a situation where the desires become ever more like fantasies, while the reality becomes one of increasing degradation.

'SWALLOWING' AND ACTS OF COMPOSURE

Living off the cruelly optimistic relation that tourism dependency often entails, means living in a state of tension between tourism's creative potential to secure the good or achieve the better life, and its undermining, upsetting character. The previous chapters of this book have shown how native locals in particular did not inhabit this state passively. They differentiated (even if with increasing difficulty, as I have argued) between times, spaces, people and tasks that stood for meaning, continuity and safety, as well as threat, absurdity and alienation. Rachid did not benefit from the same privileged position: his status as a migrant worker from abroad who had no alternative but to work in the hospitality industry, who did not own anything, and who could not count on anyone in the village by his own accounts, all contributed to his experience of stuckedness and entrapment.

Rachid may well seem like a negative, powerless pendant to his fellow Swiss and native villagers, their acts of (often covert) resistance, and their strategies of division and hierarchization (Boissevain 1996). His experience of developing a 'commodified persona' can nonetheless also be read as 'strategy', made of 'covert acts of resistance' (Bunten 2008: 381). This 'coping' with tourism was less visible than when mobilized by Swiss locals, but it helped him to survive through an everyday that had become akin to an 'impasse', as Berlant describes. Being in an impasse implies activity, although with no sense of direction, a space of skills aimed at maintaining the cruelly optimistic subject's composure, his/her impassivity and sense of self amidst the crumbling present (Berlant 2011: 4).

In Rachid's case, leaving this impasse and the hospitality industry took the most 'courage', precisely because it was disorientating. This decision opened the door to deep existential fears related to the more or less immediate future: how would he meet someone and how could he reach stability when jobless and depressed? With this act of radical change, Rachid attempted to address the cruel optimism of his life and the vicious circle in which he felt entrapped between 'working and recovering from work', as he put it. Rydzik and Anitha (2019: 884) note that quitting does not equate with 'open acts of resistance' such as protest or action aimed at changing the exploitative hospitality work relation. The latter were largely unavailable to my migrant worker interlocutors who felt particularly vulnerable given their reliance on tourism labour in the small job pool offered by the village. However, quitting, as Rachid did, radically contributes to the 'altering' of workers' lives and makes them more viable (Rydzik and Anitha 2019: 884–85).

Most common among migrant workers, however, were the smaller acts of 'resilience' that enabled them to endure 'oppressive conditions' in the everyday (ibid.). Prior to quitting his job, Rachid explained how he had been 'swallowing' (*schlucken*) the words of his peers, the humiliation, the pressure, the pain, the tensions among the work team, in order to save face in front of others – colleagues, customers or superiors. This act of swallowing was a deliberate attempt at maintaining a certain sense of self or composure in an industry and during moments that were crushing. Once, when I visited Rachid before he left his first job in the village, I had another glimpse of what such a moment of self-management implied. He stood up and handed me a pile of papers that he had carefully kept in a file, but admitted to not really understanding or caring about them. Going through the sheets, I realized these were his payslips. Looking at the papers from afar and vaguely pointing at them, Rachid explained that he did not understand how 'they' counted his hours, and he believed something was wrong as his salary fluctuated greatly every month. When he asked his employer, they cited Swiss laws and regulations he did not know about, so he decided to stop trying to understand what he was actually paid for. As he felt he would be wasting time and energy figuring this out, he just accepted being paid less than what he was worth, so as to save himself more trouble. Rachid explained that he knew he could not really leave or change anything anyway.

In this moment, I believe Rachid was wilfully trying not to understand, to care, to make himself accept. He was 'swallowing', taking in the injustice of the job and repressing a response to make it easier to go back to work the next day, without wasting unnecessary energy. In other words, Rachid was working towards an impassive attitude. For someone like him, who systematically refused to contract loans to make sure that the banks would not 'steal' the money he saved and who managed his finances meticulously, this act of resistance via impassivity surely necessitated great energy and effort. When he was diagnosed with his ulcer, Rachid directly linked his condition to so much 'swallowing' and accepting. Swallowing was a physical, mental and emotional activity that he had deployed to endure the everyday. He swallowed the bad food that he and other employees were given at the hotel, swallowed his superiors' remarks on his job, swallowed the pain he felt in front of clients as he had to put on a smile, swallowed his pride, and swallowed the anger and feeling of injustice that came with his payslips. Swallowing was simultaneously a way to endure, to navigate the difficult everyday as well as what had slowly worn down his body and mind, of which the ulcer was a result. Rachid's 'swallowing', as banal as it seems, reveals the deep incursion of his job in his intimacy, flesh and sense of self in the present. Unable to set limits to his exploitation without losing energy, and potentially his position and right to stay in Switzerland, nor to find an employer who would provide different conditions in a touristic village, Rachid's swallowing was a way to endure the everyday, when facing unfairness and the absence of alternatives. He was made to accept and continue if he ever wanted to become his own boss, to have the chance to meet someone, or to stay.

For many hospitality workers, similar strategies of endurance consisted of pushing their limits, while normalizing or naturalizing the hierarchies and relations of exploitation to find the motivation to go to work every day – a strategy described elsewhere in the hospitality work literature (McDowell, Batnitzky and Dyer 2007; Rydzik and Anitha 2019). For Andrea, who I introduced in the last chapter, refusing to give her opinion or complain was also part of the composure she had to show in a difficult present. She pushed herself to work faster and faster so she would be regarded as a 'good employee' and would not be criticized and eventually fired. After acknowledging the stress at work and the poor financial state of her household, Clara also accepted the situation by minimizing her struggles: 'We're not slaves, we're paid, so we can't really complain', she said to me, while forcing herself to go to work with a hernia, against her doctor's advice, because her boss had no alternative and she could not refuse. I suggest that all these attempts at swallowing, repressing, minimizing, normalizing, while drawing on their bodily capacities, were skills developed by those living in the impasse – to be able to work without being too disturbed or too annoyed, so as to remain professional in front of customers and bosses.

The contours of such ways of internally, covertly 'coping' with tourism (Boissevain 1996) via intimate, affective, individual and embodied self-management were those of a deeply individual practice for workers who often regarded their social surroundings with suspicion. Along my research, I was struck by the efforts deployed by the migrant hospitality employees I met to keep their lives, their joys and pains, discrete, private and secret. Rachid, who wanted me to 'tell his story', also asked me not to talk about his new job in the village, to avoid his former co-workers' jealousy, and later, to avoid feeling ashamed when he felt low. After his dismissal, he also feared that his landlord would perhaps think less of him if he knew, and think he would not be able to pay the rent, and this could eventually cost him his place. Ilana from Portugal remembered how one of her oldest friends from Portugal once went behind her back to spread rumours about her after she was congratulated in the hotel where she worked for being 'too good'. Another worker who broke his wrist while skiing asked me 'not to say anything' either – his colleagues would laugh at him, and he admitted that perhaps he had been 'reckless' to go skiing while his body was his working tool. 'Some employers formally forbid their workers to go skiing because this can cause problems for the service if you have an accident', he said; 'Not mine, but it doesn't look very good'.

CONCLUSION

In 2022 and 2023, when I came back to the village, many restaurant and hotel owners explained they were forced to close because of the lack of workforce. The guests were coming in significant numbers to the valley, but lunch service had to be cancelled and hotels that had remained open during the whole year were now forced to take vacations. Along with this, advertisements and signs recruiting workers mushroomed on the valley road. From the invisibility that Ines described, it seemed that this workforce had now suddenly become visible because it was absent. Many migrant workers, I was told by locals, had gone 'home' when the pandemic hit, never to return. The labour shortage had become another side of a (global) crisis to manage for hoteliers (Janta and Ladkin 2024), who struggled to understand where their employees had gone. From the relative ignorance of locals towards these workers when they were there, their absence now had become problematic, signifying another difficulty faced by the tourism business in recruiting 'skilled labour'.

Rachid, too, had left the valley in 2023. Pushed by the negative experiences he had had in in the village prior to and during the pandemic, along with the reduced salary he received during the lockdown, and his dreams to become his genuine self, he had relocated to another small, touristic town that offered a better quality of life. The security that the Bernese resort and his studio had offered was no longer enough. The Alpine village, its cold winter, and the lack of relationships to locals, had become unbearable. Leaving the valley once more took courage: Rachid had to accept he had to make a leap of faith into the empty future, leaving with nothing but his hopes. Still employed in a branch of the tourism industry in a town that had more than 'just' tourism to offer, he had never been closer to his dreams, he said. When I visited Rachid in his new place, I told him about the labour shortage in the valley and the struggles that some of the hoteliers and restaurants were encountering. He listened and exclaimed: 'Remember, I told you this business does not have a future!?' Rachid said that treating people badly, not offering them the possibility to 'grow old' on the job, had to have this consequence and that it had to change radically.

There was, however, no sign of such a realization in the resort. Instead, the tourism lobbyists and hoteliers of the village were determined to solve the problem by circulating employment advertisement spots, like they did for tourists, insisting on how enjoyable the village was as a holiday resort, even for a worker: 'Working where others spend their holiday' was their slogan. Time will tell whether these strategies paid off, but what they reveal is a lack of awareness of the realities experienced by people like Rachid. The hoteliers' complaints about the lack of an adequate 'skilled' workforce, before and after the pandemic, point to a form of profound misrecognition of the entrapping quality of tourism dependence for migrant hospitality workers. Accounts like Rachid's show the tendency of the hospitality industry to 'deskill' and subject workers to difficult conditions, where they feel stripped of their autonomy and opportunities to achieve a stable lifestyle, and even less able to be socially, existentially mobile. This incapacity to create trusting relationships between employers and workers, whose aging was made impossible, or even accelerated, by the industry, also led to its deterioration, as Rachid had foreseen. If I, somewhat cynically, had believed that tourism would have a future thanks to its exploitation and precarization of workers, the events following the Covid-19 crisis have proven how important it is for the hospitality industry to consider workers' well-being (Salazar 2022).

Notes

1. On the difficult working conditions in Tunisian hotels, see Saidi 2019.
2. On the importance of marriage in North Africa, and its cost for the groom particularly, see Assaad and Krafft 2015.
3. The entry of third-state nationals into the Schengen area is subject to strict visa regulations. See https://www.dfae.admin.ch/countries/tunisia/fr/home/visa/entree-ch/jusq-a-90-jours.html (last accessed 18 April 2025).
4. Or in comparison to other, more exploitative or dangerous forms of employment (Iannuzzi 2023).
5. And locals, as I have discussed in the Introduction.
6. As explained in Chapter 5, foreigners have to work in order to be given a residency permit in Switzerland. They have the right to unemployment pay, but this right is limited to two years, and can lead to the non-renewal of their permits.
7. The closest union interlocutor available to hospitality workers was in a town located one hour away from the village. However, most of the unionized workers in the region there were in the construction sector and not the hospitality industry, as one of the union's spokespersons explained. Some authors have sought explanations regarding the low presence of unions in the tourism businesses globally, proposing that the high complexity of this transnational business, which encompasses many different sorts of positions between different countries and continents, constitutes one of the reasons for this (see Baum 2007). Others have noted that 'tourism human resource issues are poorly conceptualized' on a practical and theoretical level (Liu and Wall 2006: 162).

8. This recalls Hochschild's findings among flight attendants, and her argument that service jobs often require the display of particular emotions to customers, which depend on the workers' capacity to 'manage' their own feelings. Hochschild has noted how, for service workers, emotional labour can sometimes lead to a feeling of being 'alienated from an aspect of self' (Hochschild 2003: 7).

References

Adler, Patricia A., and Peter Adler. 2004. *Paradise Laborers: Hotel Work in the Global Economy*. Ithaca, New York: Cornell University Press.

Alberti, Gabriella. 2014. 'Mobility Strategies, "Mobility Differentials" and "Transnational Exit": The Experiences of Precarious Migrants in London's Hospitality Jobs', *Work, Employment and Society* 28(6): 865–81. Doi: 10.1177/0950017014528403.

Assaad, Ragui, and Caroline Krafft. 2015. 'An Empirical Analysis of the Economics of Marriage in Egypt, Morocco, and Tunisia', in *The Oxford Handbook of Africa and Economics*.

Azcárate, Matilde Córdoba. 2020. *Stuck with Tourism: Space, Power, and Labor in Contemporary Yucatan*. Oakland: University of California Press.

Barber, Pauline Gardiner, and Winnie Lem. 2018. *Migration, Temporality, and Capitalism: Entangled Mobilities across Global Spaces*. Cham, Switzerland: Springer.

Baum, Tom. 2007. 'Human Resources in Tourism: Still Waiting for Change', *Tourism Management* 28(6): 1383–99.

Berlant, Lauren. 2011. *Cruel Optimism*. Durham, NC: Duke University Press.

Bianchi, Raoul. 2000. 'Migrant Tourist-Workers: Exploring The 'Contact Zones' of Post-Industrial Tourism', *Current Issues in Tourism* 3(2): 107–37.

Boissevain, Jeremy. 1996. *Coping with Tourists: European Reactions to Mass Tourism*, Vol. 1. New York: Berghahn Books.

Brand, Laurie A. 2006. *Citizens Abroad: Emigration and the State in the Middle East and North Africa*, Vol. 23. Cambridge: Cambridge University Press.

Brennan, Denise. 2004. *What's Love Got to Do with It?: Transnational Desires and Sex Tourism in the Dominican Republic*. Durham, NC: Duke University Press.

Bryan, Catherine. 2018. '"Wait, and While You Wait, Work": On the Reproduction of Precarious Labor in Liminal Spaces', P. Barber and W. Lem (eds), *Migration, Temporality, and Capitalism: Entangled Mobilities across Global Spaces*. Cham: Palgrave Macmillan, pp. 123–39.

Bunten, Alexis Celeste. 2008. 'Sharing Culture or Selling Out? Developing the Commodified Persona in the Heritage Industry', *American Ethnologist* 35(3): 380–95.

Çağlar, Ayse. 2018. 'Chronotopes of Migration Scholarship: Challenges of Contemporaneity and Historical Conjuncture', in P. Barber and W. Lem (eds), *Migration, Temporality, and Capitalism*. Cham: Palgrave Macmillan, pp. 21–41.

De Genova, Nicholas, and Nathalie Peutz. 2010. *The Deportation Regime: Sovereignty, Space, and the Freedom of Movement*. Durham, NC: Duke University Press.

De Haas, Hein. 2011. 'The Determinants of International Migration: Conceptualizing Policy, Origin and Destination Effects', *IMI Working Papers Series* 32: 1–35.

Duncan, Tara, David G. Scott and Tom Baum. 2013. 'The Mobilities of Hospitality Work: An Exploration of Issues and Debates', *Annals of Tourism Research* 41: 1–19. Doi: https://doi.org/10.1016/j.annals.2012.10.004.

Elgammal, Islam, and Chris Wilbert. 2013. 'Tourism Mobilities, Marriages and "Mimicking": Some Experiences of Domestic Migrant Tourist Workers in Sharm Elsheikh', *Annals of Leisure Research* 16(4): 332–47.

Griffiths, Melanie B.E. 2014. 'Out of Time: The Temporal Uncertainties of Refused Asylum Seekers and Immigration Detainees', *Journal of Ethnic and Migration Studies* 40(12): 1991–2009.

Guidetti, Giovanni, Giulio Pedrini and Luca Zamparini. 2021. 'Assessing Perceived Job Quality among Seasonal Tourism Workers: The Case of Rimini, Italy', *Tourism Economics* 27(8): 1629–49.

Guyer, Jane I. 2007. 'Prophecy and the Near Future: Thoughts on Macroeconomic, Evangelical, and Punctuated Time', *American Ethnologist* 34(3): 409–21.

Hage, Ghassan. 2005. 'A Not So Multi-Sited Ethnography of a Not So Imagined Community', *Anthropological Theory* 5(4): 463–75.

———. 2009. 'Waiting Out the Crisis: On Stuckedness and Governmentality', in Ghassan Hage (ed.), *Waiting*. Carlton, VIC: Melbourne University Press, pp. 97–106.

———. 2019. 'Afterword: Bearable Life', *Suomen Antropologi: Journal of the Finnish Anthropological Society* 44(2): 81–83.

Hazbun, Waleed. 2008. *Beaches, Ruins, Resorts: The Politics of Tourism in the Arab World*. Minneapolis: University of Minnesota Press.

Hochschild, Arlie Russell. 2003. *The Managed Heart: Commercialization of Human Feeling*. Berkeley: University of California Press.

Hodges, Matt. 2010. 'The Time of the Interval: Historicity, Modernity, and Epoch in Rural France', *American Ethnologist* 37(1): 115–31.

Iannuzzi, Francesco E. 2023. 'Local Labour Market Segmentation and Migrant Workers' Experiences: The Case of the Hotel Industry in Venice', *European Urban and Regional Studies* 30(2): 172–85. Doi: 10.1177/09697764221145357.

Janta, Hania, et al. 2011. 'Migrant Relationships and Tourism Employment', *Annals of Tourism Research* 38(4): 1322–43.

Janta, Hania, and Adele Ladkin. 2024. 'The Labour Consequences of Covid-19: Migrant Workers in Tourism and Hospitality', *Tourism Review* 79(1): 266–70.

Jordhus-Lier, David, and Anders Underthun. 2014. *A Hospitable World?: Organising Work and Workers in Hotels and Tourist Resorts*: Routledge.

———. 2017. *A Hospitable World?: Organising Work and Workers in Hotels and Tourist Resorts*. New York: Routledge.

Kalbermatter, Jacqueline. 2020. *Bleiberecht in der Gastro-Küche: Migrationspolitische Regulierungen und Arbeitsverhältnisse von Geflüchteten mit unsicherem Aufenthaltsstatus*. Zurich: Seismo.

Khosravi, Shahram. 2021. 'Afterword: Waiting, a State of Consciousness', in Christine Jacobsen, Marry-Anne Karlsen and Shahram Khosravi (eds), *Waiting and the Temporalities of Irregular Migration*. Abingdon: Routledge, p. 202.

Knox, Angela, et al. 2015. 'More Than a Feeling: Using Hotel Room Attendants to Improve Understanding of Job Quality', *The International Journal of Human Resource Management* 26(12): 1547–67. Doi: 10.1080/09585192.2014.949818.

Leite, Naomi. 2014. 'Locating Imaginaries in the Anthropology of Tourism', in Noel B. Salazaar and Nelson H.H. Graburn (eds), *Tourism Imaginaries: Anthropological Approaches*. New York: Berghahn Books, pp. 260–78.

Leitenberg, Danaé. Forthcoming. 'Navigating Tourism: Mobility, Uncertainty and Inequality in the Swiss Alps'. *Mobilities*.

Lems, Annika. 2019. 'Existential Kinetics of Movement and Stasis', *Suomen Antropologi: Journal of the Finnish Anthropological Society* 44(2): 59–80.

Liu, Abby, and Geoffrey Wall. 2006. 'Planning Tourism Employment: A Developing Country Perspective', *Tourism Management* 27(1): 159–70.

McDowell, Linda, Adina Batnitzky and Sarah Dyer. 2007. 'Division, Segmentation, and Interpellation: The Embodied Labors of Migrant Workers in a Greater London Hotel', *Economic Geography* 83(1): 1–25.

———. 2009. 'Precarious Work and Economic Migration: Emerging Immigrant Divisions of Labour in Greater London's Service Sector', *International Journal of Urban and Regional Research* 33(1): 3–25.

Narotzky, Susana, and Niko Besnier. 2014. 'Crisis, Value, and Hope: Rethinking the Economy. An Introduction to Supplement 9', *Current Anthropology* 55(S9): S4–S16. Doi: 10.1086/676327.

Perl, Gerhild. 2020. 'The Production of Illicit Lives: Racial Governmentality and Colonial Legacies across the Strait of Gibraltar', *Zeitschrift Für Ethnologie* 145: 1–20.

Pettit, Harry, and Wiebe Ruijtenberg. 2019. 'Migration as Hope and Depression: Existential Im/Mobilities in and beyond Egypt', *Mobilities* 14(5): 730–44.

Pine, Frances. 2014. 'Migration as Hope: Space, Time, and Imagining the Future', *Current Anthropology* 55(S9): S95–S104.

Ragab, Nora, Elaine McGregor and Melissa Siegel. 2013. 'Diaspora Engagement in Development: An Analysis of the Engagement of the Tunisian Diaspora in Germany and the Potentials for Cooperation'. Research report. Tokyo: UNU-MERIT.

Ramsay, Georgina. 2017. *Impossible Refuge: The Control and Constraint of Refugee Futures*. Abingdon: Routledge.

Ringel, Felix. 2018. *Back to the Postindustrial Future: An Ethnography of Germany's Fastest-Shrinking City*. New York: Berghahn Books.

Rydzik, Agnieszka, and Sundari Anitha. 2019. 'Conceptualising the Agency of Migrant Women Workers: Resilience, Reworking and Resistance', *Work, Employment and Society* 34(5): 883–99. Doi: 10.1177/0950017019881939.

Saidi, Habib. 2017. *Identité de façade et zones d'ombre. Tourisme, patrimoine et politique en Tunisie*. Paris: Pétra Editions.

———. 2019. 'Sur la biohospitalité: les hôtels comme casernes ou les trois corps du serveur', *Via Tourism Review* 16(16). Doi:10.4000/viatourism.4584.

Salazar, Noel B. 2022. 'Labour Migration and Tourism Mobilities: Time to Bring Sustainability into the Debate', *Tourism Geographies* 24(1): 141–51.

Sánchez-Rodríguez, Cristian, et al. 2022. 'Chronic Pain and Work Conditions of Hotel Housekeepers: A Descriptive Study', *International Journal of Environmental Research and Public Health* 19(6): 3383.

Seiger, Fiona-Katharina, et al. (eds). 2020. *Migration at Work: Aspirations, Imaginaries & Structures of Mobility*, Vol. 5. Leuven: Leuven University Press.

Tošić, Jelena, and Annika Lems. 2019. 'African-European Trajectories of Im/Mobility: Exploring Entanglements of Experiences, Legacies, and Regimes of Contemporary Migration', *Migration and Society* 2(1): 1–12.

Uriely, Natan. 2001. '"Travelling Workers" and "Working Tourists": Variations across the Interaction between Work and Tourism', *International Journal of Tourism Research* 3(1): 1–8.

Urry, John, and Jonas Larsen. 2011. *The Tourist Gaze 3.0*. London: Sage.

Witz, Anne, Chris Warhurst and Dennis Nickson. 2003. 'The Labour of Aesthetics and the Aesthetics of Organization', *Organization* 10(1): 33–54.

Zampoukos, Kristina, and Dimitri Ioannides. 2011. 'The Tourism Labour Conundrum: Agenda for New Research in the Geography of Hospitality Workers', *Hospitality & Society* 1(1): 25–45.

CONCLUSION

When I described the field site of my research at the beginning of my PhD in 2017, I remember the enthusiasm of many of my peers and friends as they predicted that I would be having a very enjoyable time, like they themselves have often had in an area designed to provoke experiences of relaxation, fun and amazement. I did not mind it initially, but as my fieldwork progressed I felt increasingly bothered by the assumptions of such comments, as well meaning as they were. I was less disturbed by the lack of scientific seriousness my project on tourism seemed to evoke – a complaint uttered by many tourism scholars until recently (Airey 2015: 13) – than the stark contrast such projections of peacefulness on the Alpine valley offered with the reality described by many of my interlocutors. Franziska, Moritz, Thomas, Ines and Rachid all felt, albeit differently, worried, disappointed, lonely, depressed, exhausted and overwhelmed in a place that should feel like paradise. Colourful pictures commodifying the escape, the fun and relaxation throughout the valley diametrically contradicted the everyday experiences retold by those I met who lived in the valley. Pressure, anger, fear, division and exploitation seemed to flourish in a place visited by an increasing number of guests, subjected to an accelerating but insecure market and shaped by stark differences of status between valley dwellers. Thanks to long-term fieldwork, I had accessed what seemed to be another side of tourism, made up of the people that sustained this industry with their labour (Chio 2014). As I was visiting the most paradisiac sides of the valley, the profound uncertainty of tourism, the rapid ruination and potential 'out of dateness' of the infrastructure, and the anxiety this created for those who lived off tourism, remained with me. The feeling of stepping into another dimension entering the ruined hotel I introduced at the beginning of this book never left me.

This book has been the result of my efforts to shed light on a side of tourism that has rarely been discussed for the Swiss Alps until now, namely that of tourism dependency, its experiences, politics and uneven manifestations. In this conclusion, I consider the uncertainty, paradoxes and inequality of living off tourism in the Alps, and address the political context of globalization and nativism in which they take place. I then propose a final reflection on the shape and development of tourism dependency on a global scale, shaped by the growing concerns on the effects of what has increasingly been termed 'overtourism'.

CRISES AND PROBLEMATIZING TOURISM DEPENDENCY

In retrospect, it is no surprise that hotel ruins drew me throughout my field research. As abnormal objects, contradicting the 'script of modernity' (Gupta 2018) as well as the idealizing tourism imaginaries of the region, they stood out in what I expected to be an environment shaped by success and stability. These empty, ruined or 'ghost' hotels, standing between past and future, evoked a state of crisis in the original Greek sense of the term (*κρίσις*): a moment of judgment, of verdict, of decision (Koselleck 2006). The hotels, as well as the cable car discussions and the play, generally marked a 'crucial moment that could tip the scales' (ibid.: 358) in a region that would be 'nothing without tourism'. They illustrated the 'decisive' moment the locals found themselves in: one where they had to choose between the preservation of their valley's past and the conquering of the future via speed and 'mass tourism'. Locals problematized their tourism dependency by 'temporally reasoning' with these objects and projects in particular – a typical practice of crisis moments, that 'stimulate temporal thought' (Knight and Stewart 2016: 2). As an ethnographer, I felt lucky to be able to witness what felt like an extraordinary moment, when the fate of the valley would be decided. But soon after I left the field at the end of my doctoral studies in January 2020, observing different, often simultaneous crises or 'polycrisis' (Henig and Knight 2023) unfolding in the valley, I increasingly understood that crisis was not the exception: it was the rule.

This first became apparent with the alarming crisis-talk surrounding tourism during the outbreak of and measures surrounding Covid-19, which seemed only to highlight the fragility of the pre-crisis context in which I had conducted most of my field research. In 2020, the United Nations World Tourism Organization reported that international tourism globally had 'decline[d] over 70 per cent, back to the level of thirty years ago' (UNWTO 2021). Experts evaluated the market in terms of the number of years it would take to 'recover' and come back to 'normal' (two and a half to four years according to UNWTO). Switzerland registered a decline of 40 per cent in overnight stays, also receding to the level of the 1950s, according to the

Swiss Hoteliers Association in 2021 (Bassin 2021). In the valley, the impact of months of lockdown and travel restrictions led to the JR company's most significant losses in its long history, with the its number of visitors going 'back' to those it registered in the 1980s. But if these receding figures drew on a tangible, dramatic reality of crisis in a sector that had been an exponentially booming sector since the 1970s (Eriksen 2016: 475), the fears of 'going back in time' were, as I have shown throughout the previous chapters, already very present before the exceptional break imposed by the coronavirus. Instead of pointing to moments of radical abnormality, as crisis-talk usually does (Vigh 2008; Roitman 2013; Narotzky and Besnier 2014), the coronavirus outbreak resonated with the historical preoccupations of locals to 'stay in time' with the international tourism market. In other words, such moments of crisis only revealed the intrinsic uncertainty of capitalist tourism (Bianchi and Milano 2024).

At the same time and against the assumption of crisis as a rupture of a context of normality, some people I met in the valley after the Covid-19 outbreak even reported on the appeasing quality of the crisis. Many locals retrospectively reported actually enjoying the Corona time, because of the presence of the 'good guests' – the Swiss – and the profitable business, as Switzerland had put in place relatively few measures in comparison to its neighbours. Their experiences pointed to a feeling that the pandemic times, although extraordinary, evoked more normality than business as usual. The coming 'back' of 'mass' tourists in important numbers saw a return to an upsetting normality instead: one where the wanderlust of tourists felt unmanageable again. The pandemic only seemed to highlight the abnormality of what happened before and after its impacts, namely tourism's profound uncertainty for those whose subsistence relies on it and its existential necessity for people in many places.

Finally, even the fate of the empty or ruined hotels seemed to have transitioned from one realm of crisis to another. After a bleak couple of years following the 2008 financial crisis and the 2020 pandemic, most of these hotels were bought, renovated and/or reopened from 2022 onwards – having been pulled out of their crisis-state, one assumes. They were the proof that the village had grown again, not least thanks to the new cable car project. However, a sense of crisis remained – not only because one could not 'rest on one's laurels' in the demanding tourism economy but also because the new hotels now stood for another problem: the excessive development of tourism. The resort drew on the best hotel bed occupancy in Switzerland (more than 70 per cent) but this figure was now associated with the 'density stress' (*Dichtestress*), with tourists even bothering the valley's cows, if one were to believe the media.[1] As I have shown in this book, such concerns were long present in the valley, but they seemed to reach unprecedented crisis-levels after the pandemic.

Other crises have shaped my Alpine field site since then: the 2022 energy crisis following the war in Ukraine, landslides, and the workforce crisis for example. These were the consequences of specific global events that particularly impacted the valley and its monocultural economy. However, the developments I have described above also reveal that the problematization of tourism dependency, as I encountered it in the valley, cannot be seen as the consequence of clearly defined crisis-moments, objects or projects, with a clear beginning and end that I had been fortunate to witness during fieldwork. This problematization was the consequence of a field shaped by chronic uncertainty, a field in which crisis *is* the context (Vigh 2008) because of its connection with a global situation of volatility but also because of capitalist tourism's intrinsic paradoxes as a simultaneously creative and destructive industry. Drawing on two hundred years of 'successful' touristification, my Alpine field site powerfully illustrates this reality.

CONNECTING UNEVEN EXPERIENCES OF TOURISM DEPENDENCY

Examining the experiences of many local dependents, in this book I have argued that living off tourism entails a state of ambiguity, and that looking at this dependency beyond its negative guise would help us to track the different 'modes of action', experiences and politics that make it (Ferguson 2015). The first part of this book in particular captures the unresolvable tension that characterizes tourism dependency, between romanticized visions of the Alps as a place of nostalgic longings, and capitalist endeavours to keep 'in time' with the global tourism market. This state is the social and/or individual manifestation of what Norum and Mostafanezhad (2016) have described of the tourism's 'chronopolitics', based on nostalgic or escapist, critical imaginaries of, and subjection to, existing capitalist institutions. Those who live off this business find themselves in a situation of coexistence between tourism's simultaneous purification and destruction of past-associated objects, values and identities, all for the sake of the future. The stories I have retold here reveal that the relationship between the simultaneously forward- and backward-looking activities marking tourism reliance is under constant negotiation and reformulation in a globalized, changing economy.

The objects of nostalgic yearnings incorporated into the communal debates on tourism reliance were those of a particularly remote, pre-touristic past, traditionally evoked in Alpine tourism, as well as those of a more recent touristic past, made up of European guests, synonymous with stability, meaning and profitability (Chapter 3). The futural promises evoked in the valley around the construction of the new cable car project were those of the relatively close future, but they were also inscribed in transgenerational, local ideals of security and the common good (Chapter 2). Changing in their shape and content, these temporal references speak of the paradoxical and increasingly unstable or slippery relationship between past and future that underlies capitalist tourism and structures the lives of so many (native) locals of the valley. In the inherently uncertain context of tourism, there is reason to believe that these realities will continue to manifest themselves through temporal reasoning around other temporal objects or projects in the future. Living off tourism is an art of adaptation and compromise, especially in a region that would be 'nothing' without this industry.

Paying close attention to the experience of various tourism dependents, especially in the second half of this book, has highlighted the situated nature of such experiences of temporal slipperiness. In the valley, many migrants were unable to identify or relate with the temporal frames evoked by tourism's chronopolitics in the valley. Cast as outsiders or guests under economically exploitative conditions, many felt 'stuck' in a present of existential immobility that was not necessarily disorientating but deeply disappointing and dissatisfying. As Ines and Rachid both expressed, the position and reality they inhabited were the result of being kept in a particular place, one they knew to identify as that of migrants who had 'no other choice' but to accept occupying. This entrapment was characterized by the contradictions between their individual hopes for a better life with a present of grinding conditions. Tourism mobile labour, as I have shown in the final chapter, contributed to the simultaneous emergence of these hopes and their destruction. For them, the chronopolitics of tourism was made highly personal, wrapped around their hopes for a better life on the horizon, and simultaneously standing in the way of these dreams in the corrosive present.

Slipperiness and entrapment thus constitute different expressions of tourism dependency. The long scope of temporal reasoning performed by native and Swiss locals collectively, stretching from times they had not experienced at the beginning the nineteenth century until the mid-twenty-first century, profoundly contrasts with that of migrant workers. In an individualized, enforced present, the migrants I met dreamt of a better future life. Where the former were thrown into past and future dimensions that colonized the present, the latter felt profoundly stuck in the embodied, painful present. In the valley, the migrant and native populations I conducted my research on seemed to live in different times and inhabit different 'chronotopes' or worlds – a reality that goes beyond the village or Switzerland (Çağlar 2018).

In line with what Çağlar and Glick Schiller (2018) have argued, I believe it is nonetheless crucial to attend to the ways in which these populations, as different as they were, remained linked by their experience of tourism reliance under a context of fragility and dispossession. Tracing back the conditions that allow for such differences in experiencing tourism dependency and its uncertainty, the story I have told here aims to connect these seemingly separate temporal reflexes and ontologies. What emerges from this analysis is a picture of slipperiness that is not only exclusionary in the motives evoked in local, collective endeavours to attend to the future while yearning for a romanticized past; it also sheds light on the conditions of possibility for forms of collective temporal reasoning to emerge. Slipperiness and entrapment are different but interrelated manifestations of the same reality: that of tourism dependency.

Ines and Rachid, via their presence and activities as migrant workers, allowed other locals to extract themselves from the most subjectifying, grinding, self-alienating aspects of tourism dependency. They were outsourced the most time-, energy-, life-consuming tasks of a capitalist industry based on the commodification of tourist imaginaries of escape, while increasingly reliant on speed. On the front line of service labour, migrant workers had to bear the human cost of attending the temporal tensions of tourism. The hoteliers complained about the financial consequences of a high turnover of human resources and goods implied by the market's acceleration and the guests' increasingly shorter stays, as rooms had to be thoroughly cleaned and sheets changed between guests' stays. But for workers like Rachid, it meant having to constantly adapt to new guests, new languages, new needs. For the hotel workers I followed during participant observation in one of the village's hotels, it was their bigger workload, which had to be attended to quickly and under difficult conditions, that was causing their chronic health problems. Reasoning into the distant past or the future of the industry implies a capacity to step out from these exploitative relations. Not only has the longevity of the Alpine valley's tourism industry been sustained by a succession of temporary stays of workers, especially during the *Saisonnier* regime (see Chapter 5), but it is also the awareness of this longevity, one's ability to connect with it, that can happen thanks to the incapacity of others to do so, occupied with a grinding present. The enforced present (Guyer 2007) of migrants lies in direct correlation and in relation of interdependence with the long-term projections of non-migrants.

The inequality in status and temporal projection between insiders and outsiders also suggests an uneven capacity to attend to tourism's paradoxes as a simultaneously creative and destructive industry. Tourism dependency, particularly when associated with developments that are viewed as excessive, is navigated, attended to, by those who live off it. It is 'coped' with (Boissevain 1996) and resisted in a more or less explicit manner, because of various problems that accompany the development enabled by it (Milano, Novelli and Russo 2024). Following native locals like Franziska, Moritz and Thomas has revealed how their navigation of tourism was informed by strategies of divide, avoidance and hierarchization (see Chapter 4). They ordered times, places, guests, inhabitants and tasks in the valley according to their proximity and compatibility with nostalgic ideals or their association with an alienating future. Most of my native interlocutors' efforts were aimed at living in a world that was the most meaningful, closest to the ideals of the *Heimat*, by favouring guests and places that stood for a tourism that was profitable, enjoyable and historical, and they avoided hotspots, the village centre, hotels and labour activities that were deemed alienating. In other words, they were (mostly) able to externalize the most alienating sides of tourism, and to materialize, identify, the boundary separating the good and bad sides of the industry they lived off.

For Rachid and Ines, who stood in the middle of tourist transactions, renting apartments in the middle of the village and had 'no other choice', such forms of externalization of tourism's most meaningless, unprofitable aspects were largely unavailable. Many workers told me about the difficulties they encountered when confronted with the overlap between life and work, emphasized by their chronic pain, stress or family problems. As Ines recalled of her chambermaid years, she 'did not have a life'. These difficulties were not met passively, as I have shown: the workers endeavoured to maintain their integrity, and also attempted to separate the professional from the private sphere. Thus Rachid systematically came home to eat Tunisian food he prepared for himself instead of the 'bad food' he was given at work. Sandrine, the German receptionist, although in a more privileged situation, paid particular attention to the way she ate. She explained how stress on the job even seemed to affect the way she ate, as she forced herself to slow down between each bite. And Clara explained that, in her case, the overlap between work and life was most difficult if her husband, also active in the industry, would complain about his day when he got home. She therefore established a rule forbidding any talk about work at home.

Such strategies aimed to maintain integrity, on an individual, personal or familial level, in a difficult situation. They importantly reveal the crucial sites of struggle for migrant tourism dependents: not one's *Heimat* or property, but one's body and intimacy. As such, many of my migrant worker interlocutors were unable to precisely identify the boundary between tourism's promising and threatening sides, as both intersected in the everyday, as they were part of their biographies and bodies. Their problematization of tourism dependency took a deeply intimate, personal shape, and did not find expression socially or politically: it was – is – lived, embodied and internalized on an individual level, sometimes at the level of the nuclear family.

Whereas living off tourism is marked by an inequality between the 'touring' and the 'toured ones', where the 'toured' rely on the 'goodwill of foreign visitors' (Boissevain 1996: 14), my work points to the varying positions, strategies, directions and consequences mobilized by tourism dependents. Slipperiness points to the oscillation between two identifiable and separate – even if eroding (Chapter 4) – realms of the future and the past, of un-/certainty or in-/stability, and entrapment evokes the inextricable, all-encompassing reality of precarity. As Allison describes, precarity does not limit itself to precarious work activities but spills into all aspects of one's existence. It 'is insecurity in life: material, existential, social' and 'life itself becomes a concern' (Allison 2012: 349). This was dramatically captured by Rachid's and Ines's feelings of having become someone else: while most of my native, local interlocutors yearned for different times, of greater solidarity, profitability or value, they all expressed yearning for another life, of self-actualization.

This speaks of a frequent critique of tourism as an activity fostering a sense of alienation in local communities, subjected as they are to the 'tourist gaze' or the 'tourism imaginaries' (Brennan 2004; Bunten 2008; Leite 2014, 2017). However, Rachid's and Ines's accounts, when contrasted with those of Franziska, Thomas and Moritz, underscore the importance of nuance and sensitivity in addressing these questions. They call for a careful examination of the specific sites and conditions in which alienation unfolds. Is it a 'relation of relationlessness' (Jaeggi 2014) experienced in relation to one's environment, as it is for native locals, or is it a sense of disconnection from oneself, as seen in the experiences of migrant hospitality workers?

In an industry like tourism, which remains shaped by great forms of inequality and in a global constellation of increasingly strong forms of precarization, there are strong reasons to believe that such uneven forms of dependency find expression in other global destinations. However, as noted in the Introduction, analyses inspired by the classical tourism dependency framework insisted on the negative 'impacts' of tourism, which often implied a simplistic, passive account of the community of dependents (Kaaristo 2018). Mobilizing an understanding of tourism dependency inspired by recent anthropological discussions of dependency (Ferguson 2013, 2015; Martin 2021), this book has examined how the 'toured' themselves attend to the ambiguity of tourism dependence according to their status. This perspective opens the door to unexpected realities: while the ideal of the 'free', rooted mountain dwellers in their *Heimat* continues to dominate tourism narratives and local imaginaries of the Alps, most of the natives I encountered spoke instead of experiences increasingly marked by slipperiness, disorientation and confusion. Paradoxically, those labelled as 'mobile' or 'displaced' (migrants) often felt the most emplaced – stuck or entrapped. Slipperiness therefore emerges as a mark of privilege, defined by a sense of belonging and entitlement, and the result of disappointment.

THE BUSINESS OF (MAKING) FOREIGNERS

Soon after my fieldwork, I was invited to speak to a group of Swiss anthropology students who wished to know what fieldwork 'at home' entailed. I understood the expectation of familiarity my fieldwork implied for them: I was Swiss, researching in Switzerland, in a place emblematic of the country's environment and symbols. But, contrary to such expectations, I found myself speaking about how my research in this Alpine valley had been marked by its particularly unsettling, unfamiliar nature. In the village, when speaking with Swiss, Indian, German, American, Ukrainian or Afghan nationals, I used a mixture of all the languages I know – French, German, Swiss-German, English and a little Italian – sometimes all within the same day. At the same time, I felt alien in a place where, as I was regularly reminded, I did not belong; a sense I also found among other non-native village dwellers, as well as among migrant workers, although in more radical forms. Such simultaneous existence of transnational connections and exclusionary politics of belonging in the same Alpine valley made for a reality I long struggled to understand. While many claimed that the long history of tourism had made the village a multicultural, open place, I had gathered various experiences and observations that pointed to the opposite. Living off the transnational 'business of foreigners' did not suddenly turn people, institutions or economies into cosmopolitan subjects – to the contrary. The developments of capitalist tourism did not replace more conservative, rooted forms of belonging, often associated with a past of mountain farming, but allowed for their reformulation.

This did not mean the straightforward rejection of outsiders. Virtually all native locals of the valley admitted being 'something', being able to stay in the valley thanks to the 'business of foreigners', predicated on their visits to the valley. While these explanations highlighted the economic opportunities that travellers represented for natives in an otherwise marginalized, rural region like the Alps, the presence of foreigners also contributed to the creation of their status as natives, in more symbolic or cultural terms. The Romantic outsiders' tales about Alpine environments and places crucially shaped the ideas of rooted, simple, authentic and naive mountain dwellers that still circulated in the valley during my research. These tales were, as I have suggested, still shared by both tourists and natives during my research, and they reasserted the boundary between the touring and the toured ones.

Other foreigners – migrant hospitality workers – who had contributed to the (re-) production of romanticized imaginaries of the Alps in the backstage or on the frontline of hospitality work by performing tasks that were deemed by natives to be too tedious, were made into an invisible and exploited, yet essential, part of the tourism industry. Outsourcing hospitality labour enabled natives to occupy positions and live in places they considered meaningful and authentic, providing continuity with past Alpine farming values. As one of the tourism experts on the valley put it, the village relied on this workforce from abroad to develop and run the resort, because not enough local workforce was available in the valley. Yet, this was not recognized as such, as Ines noted bitterly.

With the long-term presence of this outside foreign influence in the valley, the Bernese village was, like most Alpine valleys but especially the touristic ones, a complex borderzone (Bruner 1996) where the diverse trajectories of people crossed. All trajectories were not, however, deemed equal. Most were judged against a normative figure in the valley: that of the native, the *Einheimische*, the indigenous, original inhabitant of the valley. Cultural, historical specificities were said to make this status one of authenticity and entitlement, considering others as 'new', non-genuine, less important inhabitants. This took the form of local or regional contours, when natives affirmed the difference of other, outsider Swiss; or national, ethnic or racial contours, when they claimed the incommensurable difference of foreigners or non-Western tourists, deemed unfit, un-integrable. As I have shown, nativism in its historical, symbolic and economic facets, did not clash with the resort's globalized reality. It much rather 'ordered' it: it defined which (native) inhabitants were authentic, and it shaped hierarchizations between the (un-)wanted guests and the distribution of tourism labour between locals, according to its degrees of arduousness (Leitenberg 2025).

Ethnographically illuminating their coexistence, this book has revealed how international tourism and nativism do not contradict each other, but rather create central features of the tourism economy: they intersect in defining Romanticized tourism imaginaries and local or national ideals of belonging, but also what types of guests one wishes to receive and to whom one can easily outsource tourism labour. Tourism in the valley therefore cannot be solely described as the 'business of foreigners', based on the simple hosting of foreign travellers; it is predicated on the making and remaking of foreigners and the difference among locals *and* tourists of different backgrounds. Such politics of difference are integrated in particular regimes of hospitality dictating attitudes towards guests according to their proximity with native, Swiss or Western heritage and values, as well as defining the particular conditions for migrant reception. This is true of capitalism in general, as an economic system based on the (re-)production of difference (Kasmir and Carbonella 2014; Bear et al. 2015), but tourism, in its imaginaries, workings and experiences, cultivates difference like no other industry. If tourism dependency is shaped by a tension between a race towards the future and efforts to preserve the past, it is also shaped by an ambiguous attitude of embrace and rejection of the other, be it 'outsider', migrant hospitality worker, 'mass tourist' or historical, returning guest (*Stammgast*).

FROM SLIPPERINESS TO ENTRAPMENT AND OUTMIGRATION?

The last few decades have been shaped by increasing concerns over what many have called 'overtourism', my interlocutors included. In Barcelona, Paris and Amsterdam, many locals have complained about what seems to be an endless growth of tourism, often requesting, more or less vigorously, that tourists should 'go home' (DPA 2024) and that governments should limit the extractive power of tourism, by enforcing various forms of regulation (Milano, Novelli and Russo 2024). At the time of writing this book, Venice is the latest, and first important, global tourist destination to implement a tourist fee in order to curb mass tourism and counter the gentrifying powers of tourism. Yet, these measures have been described by critics as a 'total failure', aimed to generate 'buzz' and revenues instead of actually solving the problem (Wilson 2024). These attempts were also present in my Alpine field site, a place where tourism development has made history but also increasingly disrupted local ways of life. Attempts at giving 'back' to locals or fostering alternative types of tourism (e.g. the play *Alpine Glow*) or issuing behaviour guides for guests to mitigate the impacts of an increasing tourist presence have been, until now, minor. From an ambiguity of tourism that had been more navigable, many locals nonetheless felt increasingly powerless in the face of new, growing and accelerating forms of tourism pushed by the *Touristikers*.

Beat's case illustrates this development. Almost two years after I first met him, in 2018, I met him again. He told me that he still 'could not accept' not just the cable car's material presence but also how he had been humiliated. While he used to present himself as a proud *Ureinheimische* (genuine, native local), of local descent 'on both my mother's and my father's side', speaking the disappearing dialect and 'belonging to the mountain and the valley', Beat gravely stated that, now, being *Einheimische* 'no longer meant much'. The resentment of many of his peers remained strong, and those who sympathized with him did so mostly on a private level, fearing that they would be excluded by being associated with him, he explained. I myself, as a researcher, felt such pressure, and so I was extremely cautious when mentioning my relationship with Beat.

Beat had been excluded from the community in ways that he never thought would be possible for someone like him, epitomizing the local identity, in Switzerland, the land of freedom and private property. He now stayed at home and avoided the village centre – and not (only) because of the 'mass' tourists. He wished to avoid the hostile attitude of his peers, who were formerly opposed to the project. He even renounced going to his favourite bar in the village –taken over by an outsider in the meanwhile – where he used to sit at the regulars' table (*Stammtisch*), because 'when I enter, everyone stops talking', he stated gravely. But the distance between his home and the rest of the village, as it was now, no longer seemed much of an advantage, allowing him, like many other locals I met, to navigate tourism's ambiguity and retreat into one's *Heimat*. When Beat explained his situation, of an entrapping present marked by exclusion, I could not avoid thinking of Ines and Rachid, and how close this statement was to theirs. Beat was being ignored and he felt so bad, even at home, that he even considered leaving his home valley altogether.

Whereas Beat's case may be particular in many aspects, as an outcast, I believe it tells us something about the general direction of tourism development. This is true of rural areas like the Alps but also places that have concentrated global travellers' wanderlust for centuries, and where the middle ground of tourism seems difficult to achieve under the current constellation. Even those with access to local privilege and property, who had previously been able to navigate the slipperiness of tourism, are now at risk of experiencing a corrosive, enforced present. In Venice, this process is already well on its way, with Venetians leaving the heart of the city. And if global tourism has itself been impacted by important crises, the examples I have discussed here reveal that tourism has far from exhausted the possibility of reinventing itself, of growing and transforming. With such developments, an increasing number of new stories similar to those of Beat are likely to emerge, even in what seems to be the most stable, historical destinations, like my Swiss field site.

NOTE

1. I refrain from citing my source here in order to conform to my anonymization strategy but a journalist specifically wrote about the valley's 'density stress' in 2024.

REFERENCES

Airey, David. 2015. '40 Years of Tourism Studies – A Remarkable Story', *Tourism Recreation Research* 40(1): 6–15. Doi: 10.1080/02508281.2015.1007633.

Allison, Anne. 2012. 'Ordinary Refugees: Social Precarity and Soul in 21st Century Japan', *Anthropological Quarterly* 85(2): 345–70.

Bassin, Aline. 2021. 'La Suisse s'offre à qui voudra'. *Le Temps.* Retrieved 18 April 2025 from https://www.letemps.ch/economie/suisse-soffre-voudra?srsltid=AfmBOopR1X3-9kHejJvs26kLM076t3G_gVKavUk0th7-6dLCveh-Io1q.

Bear, Laura, et al. 2015. 'Gens: A Feminist Manifesto for the Study of Capitalism', *Society for Cultural Anthropology*, 30 March.

Bianchi, Raoul V., and Claudio Milano. 2024. 'Polycrisis and the Metamorphosis of Tourism Capitalism', *Annals of Tourism Research* 104: 103731. Doi: https://doi.org/10.1016/j.annals.2024.103731.

Boissevain, Jeremy. 1996. *Coping with Tourists: European Reactions to Mass Tourism*, Vol. 1. New York: Berghahn Books.

Brennan, Denise. 2004. *What's Love Got to Do with It?: Transnational Desires and Sex Tourism in the Dominican Republic.* Durham, NC: Duke University Press.

Bruner, Edward M. 1996. 'Tourism in the Balinese Borderzone', in Smadar Lavie and Ted Swedenburg (eds), *Displacement, Diaspora, and Geographies of Identity.* Durham, NC: Duke University Press, pp. 157–79.

Bunten, Alexis Celeste. 2008. 'Sharing Culture or Selling Out? Developing the Commodified Persona in the Heritage Industry', *American Ethnologist* 35(3): 380–95.

Çağlar, Ayse. 2018. 'Chronotopes of Migration Scholarship: Challenges of Contemporaneity and Historical Conjuncture', in P. Barber and W. Lem (eds), *Migration, Temporality and Capitalism.* Cham, Switzerland: Palgrave Macmillan.

Çağlar, Ayse, and Nina Glick Schiller. 2018. *Migrants and City-Making: Dispossession, Displacement and Urban Regeneration.* Durham, NC: Duke University Press.

Chio, Jenny T. 2014. *A Landscape of Travel: The Work of Tourism in Rural Ethnic China.* Seattle: University of Washington Press.

DPA. 2024. 'Barcelona Residents Demonstrate: "Tourists Go Home"'. *Yahoo News.* Retrieved 18 April 2025 from https://consent.yahoo.com/v2/collectConsent?sessionId=3_cc-session_82462678-e9bd-45ee-816b-e98e9c6bbb6d.

Eriksen, Thomas Hylland. 2016. *Overheating an Anthropology of Accelerated Change.* London: Pluto.

Ferguson, James. 2013. 'Declarations of Dependence: Labour, Personhood and Welfare in Southern Africa', *The Journal of the Royal Anthropological Institute* 19(2): 223–42.

———. 2015. *Give a Man a Fish: Reflections on the New Politics of Distribution.* Durham, NC: Duke University Press.

Gupta, Akhil. 2018. 'The Future in Ruins: Thoughts on the Temporality of Infrastructure', in Nikhil Anand, Akhil Gupta and Hannah Appel (eds), *The Promise of Infrastructure.* Durham, NC: Duke University Press, pp. 62–79.

Guyer, Jane I. 2007. 'Prophecy and the Near Future: Thoughts on Macroeconomic, Evangelical and Punctuated Time', *American Ethnologist* 34(3): 409–21.

Henig, David, and Daniel M. Knight. 2023. 'Polycrisis: Prompts for an Emerging Worldview', *Anthropology Today* 39(2): 3–6. Doi: https://doi.Oorg/10.1111/1467-8322.12793.

Jaeggi, Rahel. 2014. *Alienation.* New York: Columbia University Press.

Kaaristo, Maarja. 2018. 'Engaging with the Hosts and Guests: Some Methodological Reflections on the Anthropology of Tourism', in Sabina Owsianowska and Magdalena Banaszkiewicz (eds), *Anthropology of Tourism in Central and Eastern Europe: Bridging Worlds.* Lanham, MD: Rowman & Littlefield, 71–88.

Kasmir, Sharryn, and August Carbonella. 2014. *Blood and Fire: Toward a Global Anthropology of Labor*, Vol. 13. New York: Berghahn Books.

Knight, Daniel M., and Charles Stewart. 2016. 'Ethnographies of Austerity: Temporality, Crisis and Affect in Southern Europe', *History and Anthropology* 27(1): 1–18.

Koselleck, Reinhart. 2006. 'Crisis', *Journal of the History of Ideas* 67(2): 357–400.

Leite, Naomi. 2014. 'Locating Imaginaries in the Anthropology of Tourism', in Noel B. Salazaar and Nelson H.H. Graburn (eds), *Tourism Imaginaries: Anthropological Approaches*. New York: Berghahn Books, pp. 260–78.

———. 2017. *Unorthodox Kin: Portuguese Marranos and the Global Search for Belonging*. Oakland: University of California Press.

Leitenberg, Danaé. 2025. 'The Business of Foreigners: Ordering Mobilities in the Touristic Swiss Alps', in Viviane Cretton and Andrea Bosconoinik (eds), *Mountainscapes: Pathways and Mobilities in the Alps and Beyond*. New York: Berghahn Books. pp. 38–55.

Martin, Keir. 2021. 'Dependence', *Cambridge Encyclopedia of Anthropology*.

Milano, Claudio, Marina Novelli and Antonio Paolo Russo. 2024. 'Anti-Tourism Activism and the Inconvenient Truths about Mass Tourism, Touristification and Overtourism', *Tourism Geographies* 26(8): 1–25. Doi: 10.1080/14616688.2024.2391388.

Narotzky, Susana, and Niko Besnier. 2014. 'Crisis, Value and Hope: Rethinking the Economy. An Introduction to Supplement 9', *Current Anthropology* 55(S9): S4–S16. Doi: 10.1086/676327.

Norum, Roger, and Mary Mostafanezhad. 2016. 'A Chronopolitics of Tourism', *Geoforum* 77: 157–60.

Roitman, Janet. 2013. *Anti-Crisis*. Durham, NC: Duke University Press.

UNWTO. 2021. 'Impact Assessment of the COVID-19 Outbreak on International Tourism'. Retrieved 18 April 2025 from https://www.unwto.org/Impact-Assessment-of-the-Covid-19-Outbreak-on-International-Tourism.

Vigh, Henrik. 2008. 'Crisis and Chronicity: Anthropological Perspectives on Continuous Conflict and Decline', *Ethnos* 73(1): 5–24.

Wilson, Nathalie. 2024. 'Venice Entry Fee Labelled "Total Failure" with Tourists Still Flowing into the City'. *Independent*. Retrieved 18 April 2025 from https://www.independent.co.uk/travel/news-and-advice/venice-entry-fee-tourists-failure-italy-b2548563.html.

INDEX

References to notes are indicated by 'n' after the page number (33n8).

www.ingramcontent.com/pod-product-compliance
Lightning Source LLC
Chambersburg PA
CBHW072142011125
34823CB00004B/45